ELECTROMECHANICAL MOTION SYSTEMS

ELECTROMECHANICAL MOTION SYSTEMS

DESIGN AND SIMULATION

Frederick G. Moritz

FM Systems, Ohio, USA

WILEY

This edition first published 2014
© 2014 John Wiley & Sons, Ltd

Registered office
John Wiley & Sons Ltd, The Atrium, Southern Gate, Chichester, West Sussex, PO19 8SQ, United Kingdom

For details of our global editorial offices, for customer services and for information about how to apply for permission to reuse the copyright material in this book please see our website at www.wiley.com.

The right of the author to be identified as the author of this work has been asserted in accordance with the Copyright, Designs and Patents Act 1988.

Wiley also publishes its books in a variety of electronic formats. Some content that appears in print may not be available in electronic books.

Designations used by companies to distinguish their products are often claimed as trademarks. All brand names and product names used in this book are trade names, service marks, trademarks or registered trademarks of their respective owners. The publisher is not associated with any product or vendor mentioned in this book.

Limit of Liability/Disclaimer of Warranty: While the publisher and author have used their best efforts in preparing this book, they make no representations or warranties with respect to the accuracy or completeness of the contents of this book and specifically disclaim any implied warranties of merchantability or fitness for a particular purpose. It is sold on the understanding that the publisher is not engaged in rendering professional services and neither the publisher nor the author shall be liable for damages arising herefrom. If professional advice or other expert assistance is required, the services of a competent professional should be sought.

MATLAB® is a trademark of The MathWorks, Inc. and is used with permission. The MathWorks does not warrant the accuracy of the text or exercises in this book. This book's use or discussion of MATLAB® software or related products does not constitute endorsement or sponsorship by The MathWorks of a particular pedagogical approach or particular use of the MATLAB® software.

Library of Congress Cataloging-in-Publication Data

Moritz, Frederick G.
   Electromechanical motion systems : design and simulation / Frederick G. Moritz.
      pages cm
   Includes bibliographical references and index.
   ISBN 978-1-119-99274-5 (hardback)
 1. Servomechanisms.   2. Robotics.   I. Title.
   TJ214.M6576 2013
   629.8'323–dc23

                                                                    2013023514

A catalogue record for this book is available from the British Library.

ISBN: 9781119992745

Typeset in 10/12pt Times by Aptara Inc., New Delhi, India

1   2014

To my wife,
Louise,
For her enthusiastic support and encouragement

Contents

Acknowledgements

To Roger Mosciatti and Thomas Foley, my partners for over 25 years at MFM Technology; the two most inventive and dedicated engineers it has been my pleasure to know. Our efforts individually and together led to over 20 US patents and a highly successful manufacturer of motion control products and systems.

To Hans Waagen, an engineer who combines an in-depth knowledge of motion control with an ability to concisely analyze and interpret a customers' product requirements.

To Avi Telyas, founder of the Bayside Motion Group, an entrepreneur who understands and supports the engineers' contribution to the success of high tech business ventures.

To Dr. Duane Henselman, University of Maine, who has the ability to combine theory with the practical requirements of product design and manufacturing.

To Art Shoemake, for supplying many of the illustrations in this book.

To Visual Solutions, Inc., creator of the VisSim simulation software used throughout the book, for their encouragement and support in the creation of the product and system simulation diagrams.

To my wife Louise and son Peter for their able participation in checking the manuscript and proofs of the complete book.

1

Introduction

Having spent the first half of my engineering career in the design of motion control systems in computer peripheral equipment (digital tape recorders, disc memories and high speed line printers) and the second half in the design, manufacture and application of brushless motors, gearheads and slides, I have been both a user and a supplier of the various components used in high performance servo/control systems.

Time restraints in a typical one or two semester motion control/servo system course require that the text and class time devote a majority of their content to topics such as pertinent mathematics, LaPlace transform, linear system analysis, closed loop control theory and stabilization techniques. Little time can be devoted to a detailed presentation of the system components, which are typically presented in a cursory fashion only as needed to aid in the discussion of the system theory.

This book will extend such course work by providing an introduction to the technology of the major components that are used in every motion system.

A website is available (www.wiley.com/go/moritz) which will allow you to access a number of the figures in this book. You will be able to activate them, observe simulation activity, modify system parameters to observe the effect on system performance and print out the results. For example, in Figure 4.13 you can change the motor parameters (R, L, torque constant, etc.) or the load torque or the PI values to observe the effect on performance and stability. The figures that are accessible are identified throughout the book with (available at www.wiley.com/go/moritz) appended to their captions.

A system designer cannot become an expert in the design of system components. He will never fabricate his own motor or encoder, but he must understand the basics of those components in order to properly apply them and have successful discussions with his component suppliers.

I have found that one of the best sources of information about component technology is from the well written and timely application notes and white papers from component suppliers. Over the years I have accumulated a valuable file of such information and encourage everyone engaged in system design to do the same.

Electromechanical Motion Systems: Design and Simulation, First Edition. Frederick G. Moritz.
© 2014 John Wiley & Sons, Ltd. Published 2014 by John Wiley & Sons, Ltd.
Companion Website: www.wiley.com/go/moritz

Since the time I used a Bendix G15 computer in the 1960s there have been amazing advances in computer technology and application software that now allow simulation of complete electromechanical systems, including both linear and nonlinear characteristics of all the components. Throughout this book, component and system simulation is used to demonstrate the use of this technology to display system operation. A number of programs, such as MATLAB® and Simulink® by The Mathworks, MapleSim by Maplesoft, Ansys by Ansoft and VisSim by Visual Solutions are available to perform such simulations. The simulations in this book have been created with VisSim.

The book consists of the following chapters:

Chapter 2 – A brief review of servo/motion control theory, including an introduction to the use of simulation for root solving and a review of sampled data technology.

Chapter 3 – Detailed descriptions of the main components used in the fabrication of electromechanical motion control systems.

Chapter 4 – A review of the various system characteristic functions showing how they can be described and analyzed by the use of computer simulation techniques. It also contains a summary of the dynamic equations describing eight basic building blocks, which form the foundation of most motion systems.

Chapter 5 – Contains design notes and simulations of five of the many systems I helped create.

1.1   Targeted Readership

This book will be useful to:

- Students at the senior undergraduate or graduate school level to supplement their standard systems text to provide detailed component and simulation information as an aid in the design and analysis of prototype systems.
- As a reference text for graduate school students in advanced system courses involved in detailed system design and component selection.
- Practicing engineers initiating a system design requiring background information about components unfamiliar to them and/or an introduction in the use of simulation techniques.

1.2   Motion System History

A detailed and complete history of the motion control field would alone require a book larger than this one. Therefore the following just covers some highlights in the field and some of the pioneers and their accomplishments and contributions to motion control.

Since the late 1800s mathematical analysis and design of motion control has been limited to modeling the system in terms of its classical Newtonian differential equations with constant coefficients since no method existed to model nonlinearites (saturation, dead band, hysteresis, etc.) At best, linear approximations of nonlinearities have been made and designs modified once prototype test results are obtained.

This approach resulted in the mathematical representation of the system being mostly correct for small signal excitation and having errors and unpredictability for large signals.

As of 1955, Truxal summed up the status as: "The highly developed design and analysis theories for linear systems, in contrast to the rather inadequate state of the art of analysis of non-linear systems, are a natural result of the well-behaved characteristics of a linear system.Techniques for the description of non-linear systems are still in the early stage of development, with only a limited range of problems susceptible to satisfactory analysis at this time" [1].

A considerable amount of work in analysis was devoted to creating various methods of predicting stability using graphical methods showing the characteristics of the poles and zeros of the system as a result of parameter variations. Once modeled this way, a prototype system was fabricated and tested to determine what modifications to the design were necessary to account for the nonlinearities, and these were then incorporated in the calculations to more closely model the system. As the technology advanced, methods were created to analyze nonlinearities, as represented by Phase-Plane analysis (1945) (limited to second order systems) and Describing Function analysis (1950) (subject to certain system frequency characteristics).

Obviously, the reason for "designing" a system on paper is to specify the various components and predict their ability to provide proper operation before expending the time and funds to actually fabricate the system. The more accurate the "prediction", the less time it will take to complete the design.

By 1972, Tymshare Inc. of Palo Alto CA provided CSMP (Continuous System Modeling Program) which allowed simulation of systems with both linear and nonlinear blocks, operating from a main frame computer via I/O access by Teletype terminal and X/Y plotter. Bi-directional transmission of data over conventional telephone lines was slow; even a single loop simulation took as long as one hour to run and provide listed and graphical results.

Today, with the advent of the microprocessor, the PC and a wide range of simulation software, it is possible to simulate both the linear and nonlinear functionality of a complete system, combining the electrical and mechanical components and run a simulation in a matter of seconds. The "inadequate state of the art of the analysis of nonlinear systems" has been overcome.

System controllers now have rapid enough processing time such that they can perform their two main functions, that is, axis loop closure and stabilization, together with control of the functionality and interaction of all the machine axes in real time. Controllers are available that can perform such action with up to 64 axes. In addition, they can modify axis characteristics as a result of changing performance requirements; for example, software modification of stability parameters as load inertia changes.

Defining an exact time when the motion control field was started is of course not possible. It is a technology that has evolved over the past 200 years and advanced rapidly starting in the 1900s.

It is interesting that two of the earliest contemporaries in the field were James Watt (1736–1819) and Pierre-Simon, marquis de LaPlace (1749–1827). They most likely did not know each other, certainly did not consider themselves as "motion control designers" and were involved in totally opposite ends of the technology.

Watt was a "hands on" engineer, active in various aspects of the application of steam to industrial requirements. He invented an early feedback device, the fly-ball governor, which regulated the speed of an engine by controlling its supply of steam. The problem with this system was that it was often subject to oscillations; typically resolved by trial and error modifications, with no mathematical analysis involved.

LaPlace, on the other hand, was a famous French mathematician who was deeply involved in celestial mechanics, probability and field theory. He and Simeon-Denis Poisson (1781–1840) were involved with various transforms while working on probability. Although never involved in what we call motion control, one of the transforms named after him, the LaPlace Transform, is used routinely in motion control design and analysis.

Edward Routh (1831–1907) and Adolph Hurwitz (1859–1919) were also contemporaries who were specifically involved in motion control theory. Hurwitz polynomials are those which have all their zeros in the left half of the complex plane. Using this information Routh created the Routh–Hurwitz criteria, which is a simple test to determine whether a polynomial is a Hurwitz polynomial and, in turn, became the first practical test to determine the stability of a feedback control system.

Oliver Heaviside (1850–1925) was a self-educated mathematician who developed an operational calculus to solve differential equations by substituting "p" for the derivative term and thereby converting the differential equation into an algebraic equation. Although used successfully his methods were subject to much criticism by strict academics.

Harry Nyquist (1889–1976) performed work in 1932 on the use of feedback to stabilize telephone line repeater amplifiers used in long distance communications. He developed the well known Nyquist Stability Criteria which carried over into electromechanical system design. During World War II he was active in developing control systems for artillery equipment.

Harold Black (1898–1983) invented the negative feedback amplifier in 1927 which he described in a 1934 paper. He credited using Nyquist Stability Criteria to attack the stability problem created by negative feedback and discussed the effects of negative feedback, namely higher linearity, lower gain, lower distortion, higher bandwidth and reduction of nonlinearities in the forward path.

Hendrick Bode (1905–1982) extended Nyquist's work on feedback amplifier design and created the well known Bode Gain/Phase plot used to graph the gain and phase of a system as a function of frequency. This allowed the determination of the gain and phase margin of the system or how to determine its conditional stability. In 1939 he worked on military fire control systems. He authored one of the classic texts in the field, *Network Analysis and Feedback Amplifier Design*.

Walter Evans (1920–1999) developed the technique of Root Locus analysis, which shows the variation of the poles of the closed loop system with changes in the open loop gain, providing a simple means of determining how to add compensation to improve stability and performance. He authored a classic text, *Control System Dynamics*, McGraw-Hill, 1954.

Nicholas Minorsky (1885–1970) performed extensive work in analysis and design of ship steering technology. In a 1922 paper he analyzed the properties of a "three term" controller which we today call PID compensation.

Harold Hazen (1901–1980) published papers in 1934 covering the theory and design of high performance servomechanisms as a result of his work on ship shaft positioning systems.

Krylov (1879–1955) and Bogoliubov (1909–1992) developed the concept of the describing function to be used in analyzing nonlinearities. They took the approach that the system acts like a low pass or band pass filter but is limited by the transfer function being dependent on the amplitude of the input waveform. Authors of *Introduction to Nonlinear Mechanics*, Princeton University Press, 1947.

R.J. Kochenburger was the first in the United States to show the use of the describing function in his 1950 PhD thesis and in a paper,A Frequency Response Method for Analyzing and Synthesizing Contactor Servomechanisms", Trans. AIEE, Vol. 69, Part I, 1950.

H. Harris (MIT) introduced the general idea of the transfer function with respect to servomechanisms in a NRDC report in 1942.

L.A. MacColl first described the use of the phase plane to characterize a nonlinear system. *Fundamental Theory of Servomechanisms*, D. Van Nostrand, 1945.

1.3 Suggested Library for Motion System Design

There have been many books written covering various aspects of servo and control system theory and any list will be subjective and based on individual experience.

The following list is the basis for a comprehensive motion system library

- Golnaraghi, F. and Kuo, B.C. (2010) *Automatic Control Systems*, 9th edn., J. Wiley & Sons.

An up-to-date comprehensive text, covering both theory and applications.

Includes MATLAB®, ACSYS, SIMlab and Virtual Lab for on-line problem solving and application examples.

- Hanselman, D. (2003) *Brushless Permanent Magnet Motor Design*, 2nd edn, The Writers Collective.

- Bateson, R.N. (2001) *Introduction to Control System Technology*, 7th edn, Prentice-Hall.

- Doebelin, E. (1998) *System Dynamics*, CRC Press.

- Leenhouts, A. (1997) *Step Motor System Design Handbook*, 2nd edn, Litchfield Engineering Co.

A summary of the authors many years of activity in the application of stepper motors with a large number of applications.

- Levine, W.S. (ed.) (1996) *The Control Handbook*, CRC Press.

Contains an extensive summary (1500 pages, 200 authors) of modern control technology covering fundamentals, advanced methods and applications.

- Åström, K.J. and Hägglund, T. (1995) *PID Controllers*, Instrument Society of America.

In depth review of PID theory and various methods of manual and automatic tuning (Ziegler–Nichols, step response and frequency response methods).

- Kuo, B.C. and Tal, J. (1978) *Incremental Motion Control, Vol I – DC Motors and Control Systems*, SRC Publishing.
- Kuo, B.C. (1979) *Incremental Motion Control, Vol II – Step Motors and Control Systems*, SRC Publishing.

A two volume set covering DC and stepper motor driven incremental systems, with in-depth theory and a wide range of examples from actual applications

- Thaler, G.J. and Pastel, M.P. (1962) *Analysis and Design of Nonlinear Feedback Control Systems*, McGraw-Hill.

Complete coverage of the theory of the Phase-Plane and Describing Function applied to nonlinear systems including Relay (bang-bang) servomechanisms and nonlinear compensation techniques.

- Tou, J.T. (1959) *Digital and Sampled-Data Control Systems*, McGraw-Hill.

Covers the theory of sampling, the Z transform theorem and analysis, sampling frequency and root locus in the Z plane.

- Truxal, J.G. (1955) *Automatic Feedback Control System Synthesis*, McGraw-Hill.

Provides a comprehensive review of all the theory available at the time: LaPlace, RC network synthesis, Root Locus, Pole/Zero, S plane design, Z transform, Describing Function and Phase Plane analysis.

- Chestnut, H. and Mayer, R. (1951) *Servomechanism and Regulating System Design*, Vol. I & II, John Wiley & Sons.

Classic texts in the field, used for many years in senior level servo system design courses.

- James, H.M., Nichols, N.B., and Phillips, R.S. (1947) *Theory of Servomechanisms*, McGraw-Hill, 1947a volume in the MIT Radiation Laboratory Series.

An old classic covering work done during WWII on radar, gun control, torpedo and ship steering systems.

Basically treats all elements as filters with lead or derivative control and integral equalization compensation described in terms of filter theory.

No mention of nonlinearity except for a single sentence concerning backlash.

Reference

[1] Truxal, J.G. (1955) *Control System Synthesis*, McGraw-Hill, pp. 560–561.

2

Control Theory Overview

2.1 Classic Differential/Integral Equation Approach

Analysis and design of motion control systems has traditionally been accomplished by manipulation, solution and graphing of linear differential equations with constant coefficients.

Inherent in this approach is the requirement that all the system elements are time invariant and do not exhibit nonlinear characteristics (saturation, dead-band, hysteresis, etc.)

This requirement is met only by the simplest of passive components (resistors, capacitors, air core inductors, etc.) whereas all active components (motors, gearheads, amplifiers, etc.) exhibit some form of nonlinearity, at least in their large signal dynamic response. An amplifier will have a constant gain, A, such that $E_{out} = AE_{in}$ as E_{in} increases from zero until E_{out} reaches the value of the supply voltage, at which point further increases in E_{in} will not produce proportional increases in E_{out}, effectively reducing the gain and/or opening that portion of the loop in which it is located.

However, the total response of a system from the time motion is initiated until it settles stably at rest in its new commanded position consists of both linear and nonlinear operations. A system with the saturated amplifier will eventually reach null at which time linear analysis will be applicable, with some exceptions (backlash, hysteresis, stiction, etc.).

The most important achievement of modern system synthesis is the ability to combine both linear and nonlinear characteristics in a single model and determine the effect on overall performance as various parameters (linear and nonlinear) are modified without having to create laborious manual calculations and plots.

Therefore, it pays to review some of the background of the linear analysis, not as a theoretical mathematical exercise but to show its applicability to understanding system operation and as background to the introduction of computer synthesis techniques to obtain useful results without having our thinking clouded by abstract mathematics.

The general form of linear differential equation of concern is:

$$a_n \frac{d^n x(t)}{dt^n} + a_{n-1} \frac{d^{n-1} x(t)}{dt^{n-1}} + \ldots \ldots a_0 x(t) = f(t) \tag{2.1}$$

Electromechanical Motion Systems: Design and Simulation, First Edition. Frederick G. Moritz.
© 2014 John Wiley & Sons, Ltd. Published 2014 by John Wiley & Sons, Ltd.
Companion Website: www.wiley.com/go/moritz

which can be written in operator notation as:

$$\left(a_n p^n + a_{n-1} p^{n-1} + \ldots\ldots a_0\right) x(t) = f(t) \tag{2.2}$$

As a simple example, assume a rotational mass with moment of inertia J and damping constant B subjected to a constant excitation torque T. The differential equation describing the velocity (θ') and acceleration, $\left(\frac{d\theta'}{dt}\right)$ of this system is:

$$J\frac{d\theta'}{dt} + B\theta' = T \tag{2.3}$$

where:

$$J = a_1 \quad B = a_0 \quad T = f(t) \quad \theta' = x(t)$$

This equation can be rewritten as:

$$\frac{d\theta'}{dt} = \frac{T - B\theta'}{J} \tag{2.4}$$

In this form, it is easier to "think" about the system and arrive at some general conclusions about its operation:

1. Assuming zero initial conditions, since θ' is increasing with time, $\frac{d\theta'}{dt}$ is not constant.
2. When the velocity θ' reaches a value where $B\theta' = T$, $\frac{d\theta'}{dt}$ will become zero and the system will have reached a constant velocity.

Although rather simplistic, the foregoing illustrates how examining the differential equation can help to arrive at an overall understanding of a system's operation and the effect of parameter variations on performance.

For example, if the $B\theta'$ term were $B(\theta')^n$ $(n > 1)$, as it can be in a centrifuge, its effect would not be linear and the final velocity would be less than that achieved with $B\theta'$. Also, it would require higher torque to achieve the required final velocity. An example of solving this arrangement is shown later in this chapter by using simulation.

It is during an initial general examination of a systems differential/integral equation that the creative process starts to take place. It is important not to get side-tracked by pure mathematics and realize the mathematics is a means to an end. The goal is to design a stable system meeting all the specifications, using any and all design and analysis tools available.

The classic solution of this equation, using separation of variables would proceed as follows:

$$J\frac{d\theta'}{dt} = T - B\theta' \tag{2.5}$$

$$\frac{J d\theta'}{T - B\theta'} = dt \tag{2.6}$$

$$\int_0^{\theta'} \frac{d\theta'}{T - B\theta'} = \int_0^t \frac{dt}{J} \tag{2.7}$$

$$\frac{1}{-B} \ln \left(T - B\theta'\right)\Big|_0^{\theta'} = \frac{t}{J}\Big|_0^{t} \tag{2.8}$$

$$\ln \left(T - B\theta'\right)\Big|_0^{\theta'} = -Bt/J\big|_0^{t} \tag{2.9}$$

$$\ln \left(T - B\theta'\right) - \ln T = -Bt/J \tag{2.10}$$

$$\ln \frac{\left(T - B\theta'\right)}{T} = -\frac{Bt}{J} \rightarrow \frac{T - B\theta'}{T} = e^{-\frac{B}{J}t} \rightarrow \theta'(t) = \frac{T}{B}\left(1 - e^{-\frac{B}{J}t}\right) \tag{2.11}$$

This could also have been solved using the method of solving for what is variously known as the "complementary and particular" solution or the "natural and forced" [1] response or the "transient and steady-state" [2] solution.

In any case, the total solution is the sum of the two parts. For the example:

$$\text{particular, forced, steady-state solution} = \frac{T}{B}$$

$$\text{complementary, natural, transient solution} = \frac{T}{B}e^{-\frac{B}{J}t}$$

$$\text{total solution: } \theta'(t) = \frac{T}{B}\left(1 - e^{-\frac{B}{J}t}\right) \tag{2.12}$$

If Equation 2.2 is set equal to zero, that is:

$$\left(a_n p^n + a_{n-1} p^{n-1} + \ldots\ldots a_0\right) = 0 \tag{2.13}$$

it represents the general form of the denominator of the open loop transfer function of a motion system, known as the *characteristic equation*.

The roots of this equation, the values of the p terms which cause it to equal zero, determine the stability and transient response of the system. Solving the equation for the roots becomes increasingly more difficult as the order increases.

If the equation is second order:

$$ap^2 + bp + c = 0$$

it has two roots which can be determined by the well known formula:

$$p = \frac{-b \pm \sqrt{b^2 - 4ac}}{2a}$$

which will result in either two real roots or a pair of complex conjugate roots

As the order increases from quadratic to cubic to quartic to quintic, solutions become more difficult and involve experience and some intelligent guess work.

For example, a cubic must contain at least one real root plus either two more real roots or a pair of complex conjugates. Taking an initial guess at the real root and using successive synthetic divisions will eventually result in a solvable quadratic, producing the three roots.

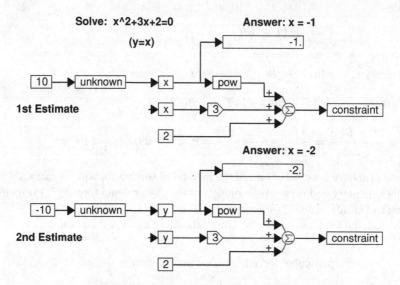

Figure 2.1   Quadratic solution by simulation

Similar techniques allow manual solving of higher order equations, but this approach is no longer necessary since computer programs and simulations (MATLAB®, VisSim) are available that provide rapid, accurate solutions.

Figures 2.1–2.4 show VisSim solution diagrams of sample quadratic and cubic equations. Note in Equations 2.1 and 2.2 the equations are "solved" by constructing them, using standard simulation blocks. The routine then starts with the initial estimates and continually calculates the equation until an answer of "0" is reached within a defined error restraint. Figures 2.3 and 2.4 use a more powerful polynomial solver.

2.2   LaPlace Transform-the S Domain

Since the solving of the differential/integral equations describing a motion system can be quite laborious, except for first and second order systems, the use of the LaPlace transform allows one to solve the equations by converting them from the time domain to the frequency domain.

The LaPlace transform, for engineering purposes, is usually stated as:

$$F(S) = \int_0^\infty e^{-St} f(t) \mathrm{d}t \qquad (2.14)$$

where S is the complex variable, $\delta + j\omega$; $\omega = 2\pi f$; f is in Hz.

Tables of direct and inverse transforms for use in LaPlace operations are readily available in control systems text books and handbooks.

The result is that calculus calculations become algebraic calculations.

The algebraic solution can be divided into a sum of common terms by partial fraction expansion and each term can then be converted back to the time domain to display the operation of the system in real time.

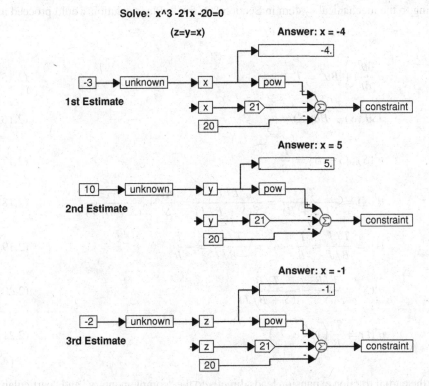

Figure 2.2 Cubic solution by simulation

Solve; X^2 +2X +2 = 0

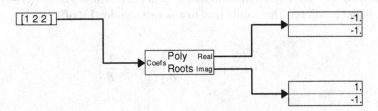

Figure 2.3 Quadratic solution by Poly Roots solver

Solve; X^3 +5X^2 +8X +6 = 0

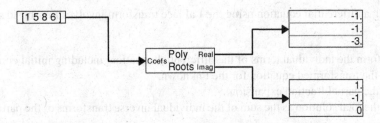

Figure 2.4 Cubic solution by Poly roots solver

Returning to the mechanical system in Section 2.1, the LaPlace solution would proceed as follows:

$$J\frac{d\theta'}{dt} + B\theta' = T \tag{2.15}$$

$$JS\theta'(S) + B\theta'(S) = \frac{T}{S} \tag{2.16}$$

$$\theta'(S)[JS + B] = \frac{T}{S} \tag{2.17}$$

$$\theta'(S) = \frac{T}{S(JS+B)} = \frac{T/J}{S(S+B/J)} = \frac{A_1}{S} + \frac{A_2}{(S+B/J)} \tag{2.18}$$

$$A_1 = \frac{T/J}{B/J} = \frac{T}{B} \quad A_2 = \frac{T/J}{-B/J} = -\frac{T}{B} \tag{2.19}$$

$$\theta'(S) = \frac{T/B}{S} - \frac{T/B}{(S+B/J)} \tag{2.20}$$

$$\theta'(t) = \frac{T}{B}\left(1 - e^{-\frac{B}{J}t}\right) \tag{2.21}$$

Note how the partial fraction expansion leads directly to the "complementary" and "particular" solutions described previously when reviewing the classical approach.

Note also, this solution assumed zero initial conditions, in which the acceleration term $\frac{d\theta'}{dt}$ transforms to $S\theta'(S)$. The exact transform, showing initial conditions is $S\theta'(S) - f(0)$, where $f(0)$ is the initial condition. This would lead to an exact solution [3] of:

$$\theta'(t) = \left[f(0) - \frac{T}{B}\right]e^{-\frac{B}{J}t} + \frac{T}{B} \tag{2.22}$$

which reduces to Equation 2.12 if $f(0) = 0$.

In other words, if this system has an initial velocity at the time the torque step is applied, the initial velocity value will exponentially decay at the same rate that the new and final value develops.

Solving a differential equation using the LaPlace transform involves four basic steps:

1. Transform the individual terms of the differential equation, including initial conditions.
2. Solve the transformed equation for the unknown.
3. Form the partial fraction expansion.
4. Form the total solution as the sum of the individual inverse transforms of the partial fraction terms.

2.3   The Transfer Function

The transfer function of any component or assembly, $H(S)$, is the ratio of the transform of its output, $F(S)_{out}$, to the transform of its input, $F(S)_{in}$.

$$H(S) = \frac{F(S)_{out}}{F(S)_{in}} \qquad (2.23)$$

In the mechanical system being evaluated:

$$H(S) = \frac{1}{SJ + B} \quad F(S)_{in} = \frac{T}{S} \quad \therefore \quad F(S)_{out} = \left[\frac{1}{SJ + B}\right]\left[\frac{T}{S}\right] \qquad (2.24)$$

A simple powerful tool provided by the concept of the transfer function, since it is a linear algebraic function and superposition holds, is the ability to manipulate and simplify combinations of transfer functions into a single unit, especially in the case of feedback loops [4].

Examples
Two transfer functions in series (Figure 2.5)

Figure 2.5   Series transfer functions combined

Two transfer functions in parallel (Figure 2.6)

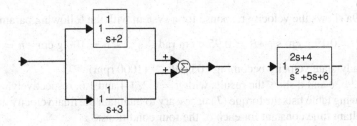

Figure 2.6   Parallel transfer functions combined

Transfer functions in parallel and series (Figure 2.7).

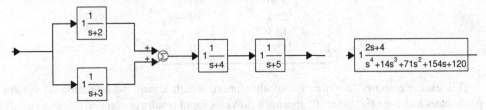

Figure 2.7   Series/parallel transfer functions combined

A second interesting aspect is to be able to manipulate transfer functions in a simulation to quickly observe the effect of component variations, both linear and nonlinear, on the operation of the system, without resorting to complex mathematics.

For example, Equation 2.7 was solved using the implicit solution:

$$\int \frac{dx}{a - bx} = \frac{1}{-b} \ln(a - bx)$$

However, as mentioned in Section 2.1, conditions could exist which would require a solution of:

$$\int \frac{dx}{a - bx^n}$$

At this point, instead of searching for a general implicit solution for this integral, and since a specific system is being designed, it is more efficient to directly simulate the transfer function in such a manner as to be able to display the velocity profile and to determine the effect the "n" factor will have on various parameters.

The transfer function can be simulated as shown in Figure 2.8 in which the viscous damping term has been isolated in order to show the effect of the "n" term.

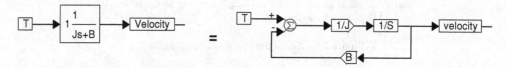

Figure 2.8   Transfer function with B term isolated

Figure 2.9a shows the velocity response for a system with the following parameters:

$$J = 0.19 \text{ g cm s}^2 \quad B = 0.95 \text{ g cm rad}^{-1} \text{ s}^{-1} \quad T = 100 \text{ g cm} \quad n = 1$$

with the steady-state velocity becoming 105 rad s^{-1} (1000 rpm)

Figures 2.9b, c and d show the results with $n = 1.2$, 1.4 and 1.6, respectively.

The following table lists the torque (T) necessary to maintain the final velocity at 105 rad s^{-1} and the resultant time constant for each of the four conditions:

n	T (g cm)	Time constant (ms)
1	100	200
1.2	254	80
1.4	645	30
1.6	1650	10

This example shows how the use of simulation, which allows manipulation of system parameters and use of a "what if" approach provides rapid results in comparison to a strictly mathematical analysis.

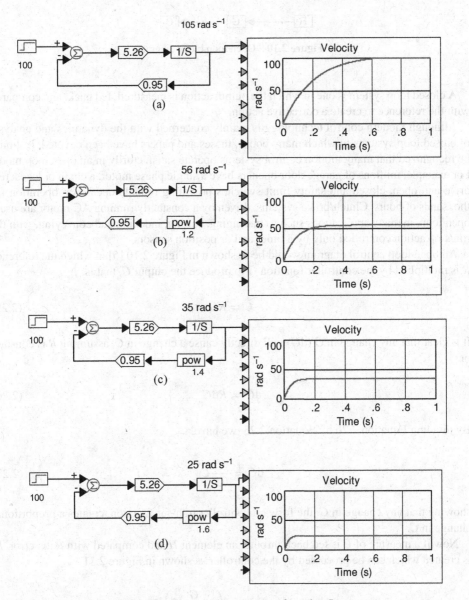

Figure 2.9   Effect on velocity response of varying B exponents

2.4   Open versus Closed Loop Control

In any motion control system, the object is to have the output of the system (position, velocity, etc.) respond to an input command and produce a desired result.

An open loop system is one in which a command (reference) is processed by the controller to the output directly without any measure of the resultant output action.

Figure 2.10   Open loop block diagram

A closed loop system is one in which the output action is measured, fed back, and compared with the reference to create a corrective action.

Although modern control technology is mainly concerned with the dynamics and analysis of closed loop systems to which many books, theses and papers have been devoted, it should be recognized that many motion control systems operate satisfactorily in an open loop mode. For example, millions of garage door openers have a single phase motor, a chain or leadscrew drive and open, closed and safety limit switches providing reliable open loop operation for thousands of hours. Clutch/brake systems, driven by a constantly running AC motor are used, open loop, in many applications such as packaging, sorting, indexing and conveying, with the run/stop action controlled only by a timer and/or position sensors.

An open loop control diagram would be as shown in Figure 2.10 [5] in which the reference R is multiplied by the controller function G to produce the output C, that is:

$$C = RG \tag{2.25}$$

It is clear that any change in G (dG) will directly cause a change in C, assuming R is constant or,

$$dC = RdG \tag{2.26}$$

By dividing Equation 2.26 by Equation 2.25, we have,

$$dC = \frac{dG}{G}C \tag{2.27}$$

showing that any change in G, the forward controller transfer function, creates a proportional change in C.

Now if a measure of C is fed back through an element H and compared with R, an error, E, is created which can be processed by the controller as shown in Figure 2.11

$$\text{resulting in } C = \left(\frac{G}{1 + GH}\right)R \tag{2.28}$$

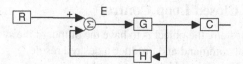

Figure 2.11   Closed loop block diagram

If we again assume a change in G (dG), then:

$$dC = \left[\frac{(1 + GH)dG - G(HdG)}{(1 + GH)^2}\right] R$$

$$dC = \left[\frac{dG}{(1 + GH)^2}\right] R$$

(2.29)

Again, dividing Equation 2.29 by Equation 2.28, we obtain:

$$dC = \left(\frac{1}{1 + GH}\right)\left(\frac{dG}{G}\right) C$$

(2.30)

showing that the effect on C by a change in the forward gain G without feedback, Equation 2.27 is reduced by the factor $\frac{1}{1+GH}$ with feedback.

Now, if we assume a change in H (dH)

$$dC = \left[\frac{(1 + GH)(0) - G(GdH)}{(1 + GH)^2}\right] R$$

$$dC = \left[\frac{-G}{(1 + GH)}\right]\left[\frac{GR}{(1 + GH)}\right] dH$$

(2.31)

substituting from Equation 2.28 and multiplying the right side by $\frac{H}{H}$ we obtain

$$dC = \left[\frac{-GH}{(1 + GH)}\right]\left[\frac{dH}{H}\right] C$$

(2.32)

Assuming that $GH \gg 1$, Equation 2.32 becomes:

$$dC = -\left[\frac{dH}{H}\right] C$$

(2.33)

which is the same form as Equation 2.27, showing that a change in the feedback creates a direct change in the output.

Conclusions:

1. In a closed loop system (a feedback system), with high loop gain at the frequencies where accurate control is desired (i.e., $C \approx R$), changes in the output caused by changes in the forward gain will be greatly reduced from those changes occurring in an open loop system.
2. However, in the closed loop system, changes in the feedback term will result in changes in the output directly proportional to the feedback changes. In other words, operation of the system (stability, accuracy, etc.) is mostly dependent on the quality, accuracy, and so on of the feedback element.

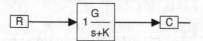

Figure 2.12   Single pole open loop diagram

2.4.1   Transient and Frequency Response

Assume we have a simple open loop system in which G has a single pole (Figure 2.12):
The response of this system to a step input $\left(\frac{R}{S}\right)$ will be:

$$C(S) = \frac{RG}{S}\left(\frac{1}{S+K}\right) \tag{2.34}$$

Using partial fraction expansion and converting back to a time function results in:

$$c(t) = \frac{RG}{K}(1 - e^{-Kt}) \tag{2.35}$$

with a time constant, TC1, of $\frac{1}{K}$

Now, if the system is modified to have feedback (Figure 2.13): the step input response will be:

$$C(S) = \left(\frac{R}{S}\right)\left[\frac{\dfrac{G}{(S+K)}}{1 + \dfrac{GH}{(S+K)}}\right] \tag{2.36}$$

resulting in:

$$c(t) = \left(\frac{RG}{K+GH}\right)\left(1 - e^{-(K+GH)t}\right) \tag{2.37}$$

where the time constant (TC2) is now $\frac{1}{K+GH}$.

The open loop time constant $\frac{1}{K}$ has been reduced to $\frac{1}{K+GH}$ which means that the transient response time has been improved (reduced).

Anticipating the content of Chapters 4 and 5 the following VisSim simulation (Figure 2.14) shows this effect, in which R is a unit step input, $G = 50$, Gain $= 5$, $H = 0.8$ and $K = 50$.

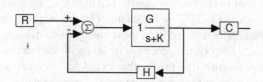

Figure 2.13   Single pole closed loop diagram

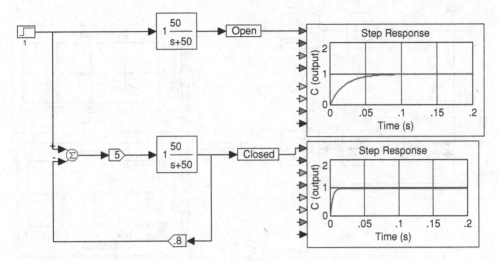

Figure 2.14   Step response of open loop versus closed loop

Note how in the open loop, $TC1 = \frac{1}{50} = 0.02$ and steady state is reached in 0.1 s, whereas in the closed loop, $TC2 = \frac{1}{50 + (250)0.8}$ and steady state is reached in 0.02 s, one fifth the open loop time.

However, also note that the forward gain has to be increased by a factor of 5 to accommodate the feedback while maintaining the input $R = 1$.

Figure 2.15 shows the frequency response of both configurations at an input frequency of 10 Hz with a peak amplitude of ± 1. Note how the closed loop output remains at ± 1 due to its lower time constant and, therefore, wider bandwidth, compared to the open loop output which has fallen to ± 0.6.

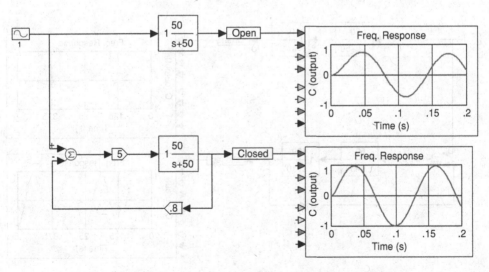

Figure 2.15   Frequency response (10 Hz) of open versus closed loop

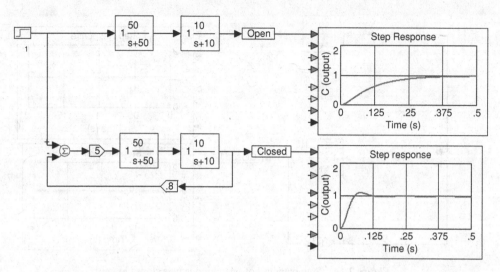

Figure 2.16   Step response of two pole open versus closed loop

Next, if we expand the system to have two break points (time constants), indicative of the electrical and mechanical time constants of a velocity servo, the transient and frequency responses are shown in Figures 2.16 and 2.17.

The open loop system has a transfer function of:

$$\frac{C(S)}{R(S)} = \frac{500}{S^2 + 60S + 500} \quad \text{with} \quad \omega_n = \sqrt{500} = 22.4 \text{ rad s}^{-1} = 3.6 \text{ Hz}$$

$$\text{and} \quad \zeta = 1.34$$

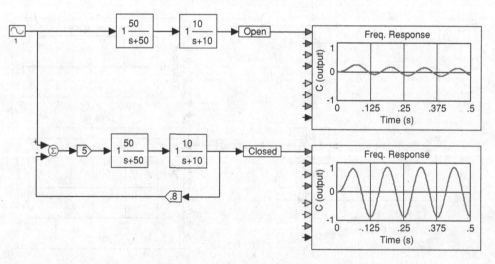

Figure 2.17   Frequency response (10 Hz) of two pole open versus closed loop

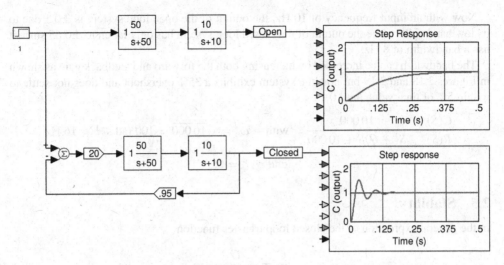

Figure 2.18   Repeat of Figure 2.16 with increased gains

and is an overdamped system with narrow bandwidth compared to the closed loop system which is now a typical second order system, where:

$$\frac{C(S)}{R(S)} = \frac{2500}{S^2 + 60S + 2500} \quad \text{with} \quad \omega_n = \sqrt{2500} = 50 \text{ rad s}^{-1} = 8 \text{ Hz}$$

$$\text{and} \quad \zeta = 0.6$$

The step response of the closed loop system exhibits only a 10% overshoot and settles to within 5% of the final value in 0.1 s.

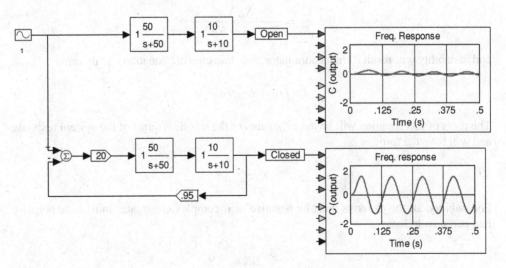

Figure 2.19   Repeat of Figure 2.17 with increased gains

Now with an input frequency of 10 Hz, the output of the open loop system is ±0.1 due to its low bandwidth while the output of the closed loop system has only fallen to ±0.65 since it has a bandwidth of 8 Hz.

The bandwidth can be increased by increasing both the forward and feedback gain as shown in Figures 2.18 and 2.19 but now the system exhibits a 35% overshoot and does not settle to within 5% of final value until 0.13 s.

$$\frac{C(S)}{R(S)} = \frac{10\,000}{S^2 + 60S + 10\,000} \quad \text{with} \quad \omega_n = \sqrt{10\,000} = 100 \text{ rad s}^{-1} = 16 \text{ Hz}$$

$$\text{and} \quad \zeta = 0.3$$

2.5 Stability

In the general expression of the closed loop transfer function

$$\frac{C}{R} = \frac{G}{1 + GH}$$

any combination of G and H for which $1 + GH = 0$ will cause $\frac{C}{R}$ to be infinite.

Specifically, if the magnitude of GH is equal to or greater than 1 at a frequency at which the phase angle of GH is equal to $-180°$, the system will be unstable.

The terms G and H each consist of polynomials with a numerator and denominator such that:

$$G \to \frac{N_G}{D_G} \quad H \to \frac{N_H}{D_H}$$

then:

$$\frac{C}{R} = \frac{N_G N_H}{D_G D_H + N_G N_H}$$

and instability will result if the denominator, the characteristic equation equals zero:

$$D_G D_H + N_G N_H = 0$$

The roots of this equation will be the exponents of the transient terms of the system response and will have the form:

$$K_n e^{p_n t}$$

For stability, all the p_n terms must be negative or if complex conjugates must have negative real parts, such that:

$$K_n e^{p_n t}\big|_{t \to \infty} = 0$$

meaning that all exponential terms eventually decay to zero.

If any p_n term is positive, instability will result, with the system exhibiting a constant sinusoidal oscillation or a violent oscillation between positive and negative values determined by system power limits, nonlinearity, and so on.

Determining the p_n term for second or third order characteristic equations is fairly straight-forward. For higher order equations, the Routh/Hurwitz technique was developed to determine the existence of roots with positive real parts. However, with modern software algorithms, such as ROOTS in MATLAB® or IMPLICIT SOLVE in VisSim (see Figures 2.1–2.2) it is possible to solve an equation of any order.

What is also of interest is the subject of relative stability. A system with a damping factor of 0.3 is stable, as is one with a damping factor of 0.7. However, the transient terms of the one with 0.3 damping will take much longer to decay than the one with the 0.7 damping.

Much of the work in the use of Bode plots, Nyquist diagrams, root locus plots and even phase plane diagrams is to determine how to modify or compensate systems to achieve required performance together with reasonable relative stability.

Relative stability is usually defined in terms of the *gain margin and phase margin* of the system, especially with reference to the open loop Bode plot of the system.

The *gain margin* is the amount the gain falls less than 0 db when the phase is $-180°$.

The *phase margin* is the amount the phase is greater than $-180°$ when the gain is 0 db.

What these margins should be for "acceptable" performance is subject to the individual designers preferences and the demands of the system.

But experience has shown that in general a gain margin of -6 db and a phase margin of $40°$ results in acceptable performance [6].

2.6   Basic Mechanical and Electrical Systems

2.6.1   Equations and Constants

The linear dynamic operation of any DC (brush or brushless) motor driven electromechanical system can be described by the following equations;

$$E_a = K_e\theta' + I_a R_a + L_a I_a' \tag{2.38}$$

$$T = J\theta'' + B\theta' + T_l \tag{2.39}$$

$$T = K_t I_a \tag{2.40}$$

Equation 2.38 describes the current (I_a) in the motor as a function of the applied voltage (E_a), the BEMF ($K_e\theta'$) and the motor resistance (R_a) and inductance (L_a).

Equation 2.39 describes the acceleration and velocity of the system as a function of the applied torque (T), the load torque (T_l) and the system inertia (J) and viscous damping (B).

Equation 2.40 ties Equations 2.38 and 2.39 together, calculating the torque (T) generated by the motor current (I_a).

In order to apply these equations in a system simulation, the motor and load parameters must be researched and/or calculated or measured.

The motor electrical and generated parameters:

Voltage constant K_e: V rad $^{-1}$ s^{-1} or V/1000 rpm

Torque constant K_t: N m A^{-1} or oz-in A^{-1}

Resistance R_a: Ω

Inductance L_a: mH

are typically available from the manufacturers data sheet. Also, since both K_e and K_t are created by the interaction of two magnetic fields (see Section 3.1) they are directly related and if only one is known, the other can be calculated as follows:

$$K_e \left(\text{V rad}^{-1} \text{ s}^{-1} \right) = K_t \left(\text{N m A}^{-1} \right)$$

or

$$K_e \left(\text{V}/1000 \text{ rpm} \right) = 0.74 K_t \left(\text{oz-in A}^{-1} \right)$$

The load parameters:

Inertia J: g cm s^2 or oz-in-s^2

Load Torque T_l: g-cm or oz-in

Viscous Damping B: g-cm rad^{-1} s^{-1} or oz-in rad$^-$ s^{-1}

can be obtained by data sheet research or calculated as follows:

Inertia J

The concept of inertia and its effect on system performance and torque requirements is often confusing to a first time system designer. Its confusion with weight can lead to wrong calculations and misleading response predictions. A more detailed review of inertia and its effect on system performance is discussed in Section 4.5.

The total system inertia is the sum of the motor inertia plus all the additional inertias reflected to the motor shaft. Motor and gearhead inertias are typically available from manufacturers' data sheets. Lead screw and pulley inertias are not always available, but can be approximately calculated using the formulas for disc and cylinder inertias listed in Section 6.2.

The load inertia is the most difficult to obtain, and in many cases is the most important and largest inertia, since it is typically composed of an irregularly shaped object made of various materials. Therefore, the load must initially be approximated as an assembly of blocks, cylinders, and so on to arrive at a value to be used for initial simulations. Once a prototype system is available, the actual load inertia can be measured by testing system response and then used to update the simulation.

Load Torque T_l

Load torque is typically composed of the bearing torques of the total structure plus any actual uni-directional or bi-directional torques created by the load. Initially, bearing torques can be assumed to be in the range of 36 to 144 g cm per bearing. If a lead screw is being used, the anti-backlash adjustment of the nut assembly can create significant friction torque, which is usually specified in the data sheet. In addition to "normal" or Coulomb friction torque, stiction (or "breakaway") torque must be accounted for and can be simulated (see Section 4.8.2).

Viscous Damping B

The viscous damping constant is important in determining both system static and dynamic response and yet is hardly ever specified in component data sheets. It is not designed into a product, but results as a by-product. For example, a gearhead will be designed to have a certain gear ratio, torque rating, maximum speed, and so on and be fabricated with a certain amount and grade of lubricant. As a result, it will have a certain value (as measured) of viscous damping. This could be altered by minor changes in the design and lubricant, but the viscous damping is not considered to be a prime, well controlled parameter of the gearhead. This is also true of motors, lead screws, ball screws, belts and pulleys, and so on. It is also true of the primary load. The load is designed to achieve a certain performance, not to exhibit a predefined viscous damping effect, and yet the total assembly will have viscous damping which will contribute to system performance. Also, in high speed systems or systems moving large air loads, such as centrifuges, B itself can be variable, proportional to the second or third power of velocity.

In addition, text books on servo and automatic control theory, when discussing system examples, rarely, if ever discuss the details of the damping constant but simply list it as a given value.

The theoretical effect of the damping constant was reviewed in Section 2.1, showing that the final, steady-state velocity of the example is:

$$\theta' = \frac{T}{B} \tag{2.41}$$

So, in a mechanical assembly, operating in an open loop condition, with a fixed value of applied torque, the viscous damping will determine the final steady state velocity of the load. Without viscous damping, the load could, theoretically, accelerate to an infinite velocity.

This is analogous to the current in an electrical circuit, which in the steady state is determined by the resistance and in which the current would rise toward infinity as the resistance drops toward zero.

Another effect of viscous damping is shown in the characteristic equation of a second order motor-driven system (see also Sections 2.1–2.4 and the examples in Chapter 5).

$$S^2 + \frac{(J R_a + B L_a)}{J L_a} S + \frac{(K_e K_t + B R_a)}{J L_a} \tag{2.42}$$

In this expression $\frac{(K_e K_t + B R_a)}{J L_a}$ is ω_n^2, the square of the system natural frequency and $\frac{(J R_a + B L_a)}{J L_a}$ is $2\zeta \omega_n$ where ζ is the system damping factor. Note that the system damping factor is a function of B, that is, the viscous damping constant is not the same as the system damping factor!

The following describes two methods that can be used to measure the viscous damping constant.

2.6.2   Power Test

If the motor is operated, no load, at a very low velocity while the input power from the power supply is recorded, this power will mainly be determined by the motor bearing friction.

Secondly, if the motor is then operated at a higher velocity, the input power will then be determined by the bearing friction plus the viscous friction, which will allow the viscous constant, B, to be calculated.

Example

A motor running at 500 rpm required 0.6 W of power, equal to 8×10^{-4} HP.

From this the bearing torque is calculated to be 117 g cm.

When running at 7000 rpm, the motor required 17.7 W.

Therefore the viscous damping power is: $17.7 - 0.6 = 17.1$ W, equal to 2.3×10^{-2} HP

The viscous damping torque can then be calculated to be 238 g cm at 7000 rpm.

From this: $B = 0.32$ g cm rad^{-1} s^{-1}

Note that this method is approximate. A more exact procedure would be to also subtract the $I^2 R$ and core losses from the input power. This would result in a lower value for B.

2.6.3   Retardation Test

If the motor is operated at some steady-state speed and then disconnected from the supply, it will coast to a stop under the influence of both the bearing torque (a linear effect) and the viscous damping (an exponential effect).

The velocity will decay to zero according to:

$$\theta'(t) = \left(\frac{T_l}{B} + \theta_0' \right) e^{-Bt/J} - \frac{T_l}{B} \tag{2.43}$$

Also, if $B = 0$,

$$\theta'(t) = \theta_0' - \frac{T_l t}{J} \tag{2.44}$$

Figure 2.20 shows the results of a retardation test performed on a motor with the following parameters:

$$J = 38.2 \text{ g cm s}^2 \qquad T_l = 828 \text{ g cm} \qquad V_0 = 78 \text{ rad s}^{-1}$$
$$\text{At } 0.5 \text{ s}, \ V = 56 \text{ rad s}^{-1}.$$

These data were then used to perform a VisSim simulation calculation to solve for B, as shown in Figure 2.21 resulting in $B = 12.8$ g cm rad^{-1} s^{-1}.

Figure 2.20 also shows a plot of the velocity decay that would occur if $B = 0$, demonstrating the initial effect of B and how its effect tapers off as the velocity decays and bearing friction predominates at the end of the trace.

Once all of the system parameters are known, the system can be simulated using VisSim simulation software, as shown in Figure 2.22.

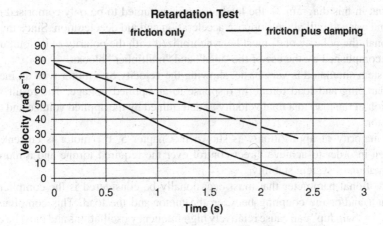

Figure 2.20   Retardation test result

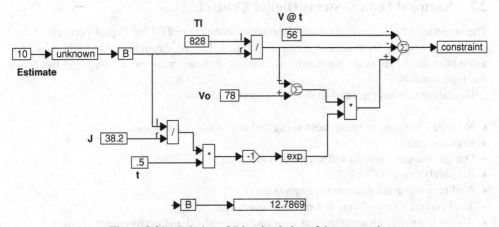

Figure 2.21   Solution of *B* by simulation of decay equation

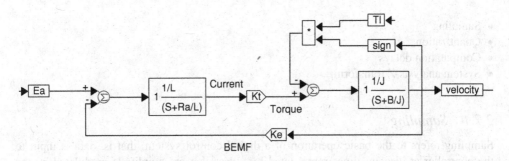

Figure 2.22   Basic motor block diagram with bearing load torque

Note that in this diagram T_l, the load torque, is assumed to be only comprised of bearing friction and as such will always oppose acceleration and aid deceleration. Since the system is bidirectional, the polarity of T_l must be synchronized with the polarity of the output velocity. This is accomplished by the use of the "sign" and "multiplier" blocks.

The system simulated is very basic, showing the velocity of a motor with an inertial load, viscous damping and load torque in response to an applied voltage. The result will be an acceleration of the load to a final velocity as determined by the applied voltage and the BEMF of the motor.

In the majority of applications, as shown in Chapter 5, the motor is current, not voltage, driven in order to achieve direct control over the required torque and a more efficient implementation of system stability.

One additional parameter that must occasionally be considered is the compliance of the driven shaft and/or the coupling between the motor and the load. This compliance (K) or "stiffness" or "windup" can cause relatively high frequency oscillations and must be eliminated using filter and compensation techniques. See Chapter 4, Section 4.6 for an example of the action of compliance in a ball screw system.

2.7   Sampled Data Systems/Digital Control

The majority of electromechanical systems are now controlled by digital computer-based controllers, ranging from simple single-chip-based designs through one to eight axis PCB assemblies to large scale multiaxis assemblies forming part of a complete bus-based machine.controller.

Digital control has replaced analog for a number of reasons:

- No drift – immune to component aging and environment changes.
- High accuracy.
- Design changes made in software, not hardware.
- Relatively low cost.
- Ability to store and retrieve many programs.
- Total control – servo loops and machine operation.
- Ability to modify loop parameters to accommodate load and performance changes.

There are four main features of all sampled data systems:

- Sampling
- Quantization
- Computation delays
- System analysis-Z transform.

2.7.1   Sampling

Sampling refers to the basic operation of a digital control system, that is, data is input to the controller at discrete times represented by pulses that are amplitude modulated by the incoming signal.

Shannon has shown that "if a signal is sampled instantaneously at a constant rate equal to twice the highest signal frequency, the samples contain all the information in the original signal" [7].

Although this theory laid the foundation for much of modern communication technology, when applied to motion control it is somewhat lacking.

Assume a 100 Hz sine wave is transmitted through a sample and hold circuit at four different sampling rates, as shown simulated in Figure 2.23.

When sampling occurs at a rate 100 times per signal cycle, or a sample rate of 10 KHz, the sample and hold output is a "perfect" representation of the input with only a small amplitude perturbation at the sampling rate.

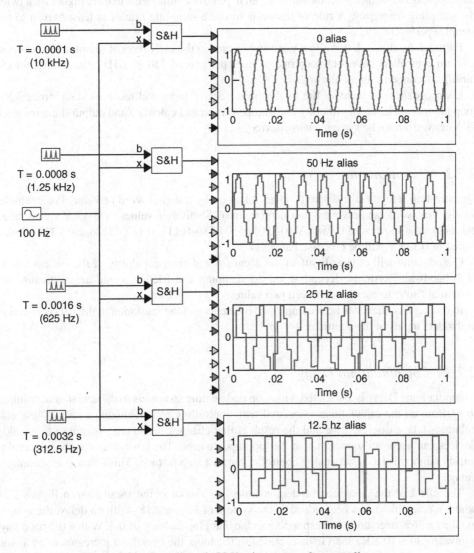

Figure 2.23 Sampling of 100 Hz sine wave at four sampling rates

At a sampling rate 12.5 times per signal cycle, or a sampling rate of 1.25 KHz an amplitude modulated frequency lower than the 100 Hz input signal appears in the sample and hold output. This signal is a 50 Hz signal at approximately 2.5% peak to peak.

At a sampling rate 6.25 times per signal cycle, or a sampling rate of 625 Hz an amplitude modulated frequency of 25 Hz develops at 10% peak to peak.

Finally, at a sampling rate of 3.125 times per signal cycle, or a sampling rate of 312.5 Hz an amplitude modulated frequency of 12.5 Hz develops at 40% peak to peak.

This effect is called "*aliasing*" in which "one or more frequency components are added to the original signal" [8].

If this effect isof concern, then the only way to eliminate or minimize it would be to increase the sampling rate and/or filter the input signal to remove components that are higher than twice the sampling frequency. A rule of thumb is to use a sampling rate of at least "seven to ten samples per cycle" [9].

Fortunately, although this effect was troublesome in the early days of sampled data control, modern controllers have total loop update rates per axis of 5 to 20 KHz, eliminating, for the most part, concerns about aliasing.

Loop update time includes the total time to receive input and feedback data, process the loop algorithms including filtering and compensation and calculate and output the error via a D/A conversion to the system analog devices.

2.7.2   Quantization

Quantization refers to the discrete values taken on by a digital word or value. For example, an eight bit word can only exist, or quantize, into 256 discrete values. A 0–15 V signal range can only exist in steps of 0.0586 V (0–0.0586, 0.0586–0.117, 0.117–0.176, etc.) Therefore a signal of 0.125 V will be processed as 0.117 V.

Quantization will obviously affect the accuracy and error capability of the system and in high bandwidth systems, depending on the damping and friction, could actually cause the system at "null" to oscillate between two values.

Reducing this effect can be accomplished by increasing the resolution of the A/D conversion at the data inputs of the controller.

2.7.3   Computational Delay

Computational Delay is associated with loop update time and refers to the actual time required to perform all the calculations associated with controlling and stabilizing a loop. These calculations take a discrete time and the result is that changes in the error output to the analog devices can occur no sooner than at the loop update time. The error does not change continuously (smoothly) as in an analog controller but in a step fashion. This effect is essentially a delay.

The effect of the delay can be demonstrated, as shown in the simulation in Figure 2.24 where a delay block has been added to the system of Figure 2.18. With no delay, the system exhibits a 36% overshoot, corresponding to the damping factor ζ of 0.3. With a 0.1 ms delay, equivalent to a 10 KHz loop update, added to the loop, the overshoot increases to 38% and with a 1 ms delay (a 1 KHz loop update) the overshoot increases to 46%.

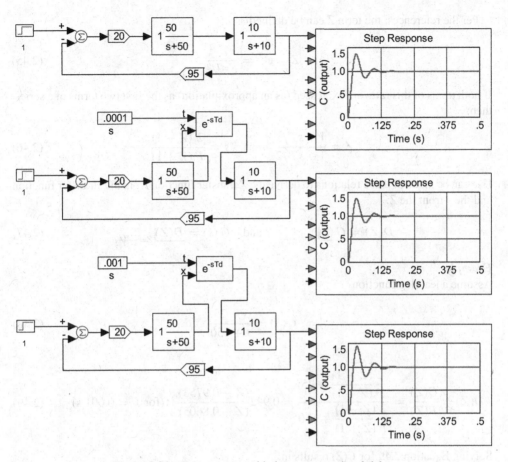

Figure 2.24 Step response with three computational delays

Again, the higher the loop update rate, the lower will be the negative effect of the computational delay on system performance and stability.

2.7.4   System Analysis

Analysis of sampled data systems makes use of the Z transform, which is essentially the LaPlace transform of a sequence of pulses.

The Z transform, like the LaPlace transform, is a linear transform and, therefore, can be manipulated like the LaPlace. Stability, gain and phase margin, and so on can be evaluated in the Z plane. A detailed discussion of "Z"-based mathematics is outside the range of this book but can be found in a number of references [10–12].

Of interest is the ability to convert an "s" transfer function into a "Z" transfer function using one of several methods, the most popular of which is the Bilinear.Transform. All of the conversion methods make use of approximations in order to simplify analysis.

Per the references, the term Z can be defined as:

$$Z = e^{sT} = \frac{e^{sT/2}}{e^{-sT/2}} \tag{2.45}$$

If both terms of this ratio are expressed (as an approximation) as the first two terms of a series, then:

$$Z = \frac{1 + sT/2}{1 - sT/2} \quad \text{and} \quad s = \frac{2\,(Z - 1)}{T\,(Z + 1)} \tag{2.46}$$

Use can be made of these relations to derive the Z transfer function from the s transfer function and the s from the Z:

$$D(Z) = \underline{G\,(s)}\Big|_{s = \frac{2}{T}\frac{(Z-1)}{(Z+1)}} \quad \text{and} \quad G\,(s) = \underline{D(Z)}\Big|_{Z = \frac{1+sT/2}{1-sT/2}} \tag{2.47}$$

Example
Assume a lead/lag function

$$G(S) = \frac{s + 25}{s + 150} \tag{2.48}$$

$$D(Z) = \frac{C(Z)}{R(Z)} = \frac{\dfrac{2\,(Z - 1)}{T\,(Z + 1)} + 25}{\dfrac{2\,(Z - 1)}{T\,(Z + 1)} + 150} = 0.942\frac{(Z - 0.9753)}{(Z - 0.8605)} \quad \text{(for } T = 0.001 \text{ s)} \tag{2.49}$$

Solving Equation 2.49 for $C(Z)$ results in:

$$C(Z) = 0.8605\frac{C(Z)}{Z} + 0.942R(Z) - 0.9187\frac{R(Z)}{Z} \tag{2.50}$$

Recognizing that $1/Z$ represents a delay, this result shows the calculation the processor must perform to emulate the lead/lag function, that is, the new output equals the sum of 0.8605 times the previous output plus 0.942 times the present input minus 0.9187 times the previous input.

An interesting aspect of sampled data design can be shown by synthesizing the Z transform of an integrator with a step input for various sampling times, as shown in Figure 2.25.

In Figure 2.25 sampling is at (a) 20 (b) 100 and (c)1000 Hz. Figure 2.25a also has the output of a pure (analog) integrator displayed for comparison.

This shows in graphical form the rather obvious conclusion that as T approaches zero, that is, the number of sampling pulses approaches infinity, a sampled data system approaches becoming an analog system.

Since any component which has an s- based transfer function can have it converted into a Z-based transfer function, a complete analysis can be performed in Z or in s.

The decision of which approach to use is mainly dependent on the experience and preference of the designer. One important use of the Z transform is to aid in developing the computer

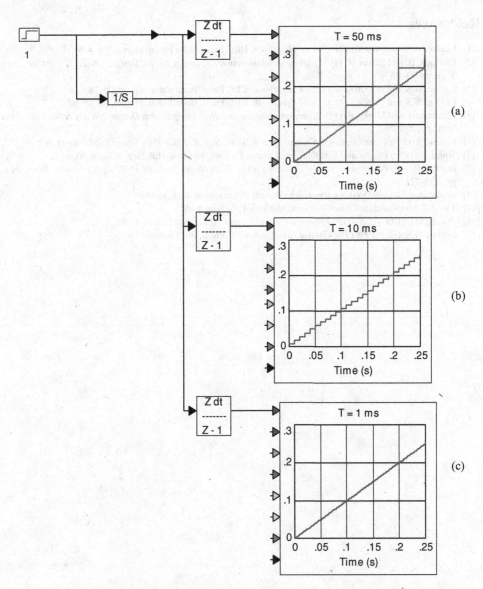

Figure 2.25   Step response of Z integrator at three sampling frequencies

algorithms that will control the system. Since the system designer will not be the controller designer, just like he will not be a motor designer but a motor user, the Z technology will not be used directly by him, but he should have knowledge of it to understand its use in his system, just as he has to know how a motor works and its capability in order to apply it properly.

In the majority of applications, s plane design and analysis together with proper selection of processors or computers that have loop update rates of 5 KHz or higher will be an efficient design approach.

References

[1] Levine, W.S. (ed.) (1996) *The Control Handbook*, CRC Press, Boca Raton, Florida, pp. 4–8.

[2] Chestnut, H. and Mayer, R. (1951) *Servomechanisms and Regulating System Design*, J.Wiley & Sons Inc., New York, pp. 33–35.

[3] Levine, W.S. (ed.) (1996) *The Control Handbook*, CRC Press, Boca Raton, Florida, p. 28.

[4] Levine, W.S. (ed.) (1996) *The Control Handbook*, CRC Press, Boca Raton, Florida, pp. 86–92.

[5] Chestnut, H. and Mayer, R. (1951) *Servomechanisms and Regulating System Design*, J.Wiley & Sons Inc., New York, pp. 200–202.

[6] Batson, R. (1996) *Introduction to Control System Technology*, Prentice Hall, Upper Saddle River, N.J., p. 572.

[7] Tou, J.T. (1959) *Digital and Sampled-data Control Systems*, McGraw-Hill, New York, p. 80.

[8] Batson, R. (1996) *Introduction to Control System Technology*, Prentice Hall, Upper Saddle River, N.J., pp. 220–221.

[9] Doebelin, E.O. (1998) *System Dynamics*, Marcell Dekker, New York, p. 187.

[10] Tou, J.T. *Sampled-data Control Systems*, McGraw-Hill, New York,

[11] Truxal, J.G. (1955) *Automatic Feedback Control System Synthesis*, McGraw-Hill.

[12] Levine, W.S. (ed.) (1996) *The Control Handbook*, CRC Press, Boca Raton, Florida,

3

System Components

3.1 Motors and Amplifiers

Traditionally, motors, in a very broad sense, have been categorized by the type of power applied to them, that is, AC or DC. In addition they can operate independent of electronics; their basic operation was, and still is to a large extent, self-contained.

AC motors of all types and DC brush motors will run by simply supplying them with the appropriate power; AC single phase, AC three phase, DC. In addition, AC has typically meant sinusoidal.

However gray areas have always existed with regard to the AC/DC definition of a motor as exemplified by the series motor which will operate on both AC and DC power and is, therefore, better known as a Universal motor.

The development of the brushless DC motor and the stepper motor has created a new class of motors that cannot operate without associated drive electronics. Even three phase variable speed/variable torque AC induction motors can only perform with a companion electronic drive assembly.

DC motors, both brush and brushless, use shaft position information to perform the basic function of commutation; brush mechanically (Section 3.1.2) and brushless electronically (Section 3.1.4). But, the brushless motor has a rotor with permanent magnets, a three phase stator and, in the version with sinusoidal BEMF, is typically driven by an amplifier with sinusoidal output power. A synchronous AC motor also has a rotor with permanent magnets and a three phase stator but is driven from a three phase supply, independent of shaft position.

Schematically, the two motors are identical. They are both brushless but differ in the method power is supplied to them. The synchronous motor is obviously an AC motor but the brushless motor is DC by virtue of its shaft position sensed, electronically controlled commutation, but AC by virtue of the form of the drive power.

The stepper motor is brushless and termed an AC motor but in full step mode is driven by sequentially applied bi-polar current pulses, neither DC nor sinusoidal AC, but becomes pseudo-sine driven in the high resolution microstepping mode.

Electromechanical Motion Systems: Design and Simulation, First Edition. Frederick G. Moritz.
© 2014 John Wiley & Sons, Ltd. Published 2014 by John Wiley & Sons, Ltd.
Companion Website: www.wiley.com/go/moritz

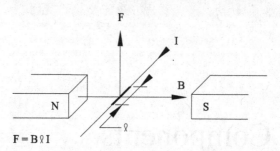

Figure 3.1 Force on current carrying conductor in magnetic field

This section reviews the four most popular motors used in electromechanical control systems, detailing modes of operation, limitations and application information for:

- Brush motors
- Brushless motors
- Stepper motors
- Induction motors.

3.1.1   Review of Motor Theory

The work of Jean Baptiste Biot (1774–1862), and Felix Savart (1791–1841), formulated the relationship between current in a conductor and the magnetic field surrounding it, which led to Hedrick Antoon Lorentz (1853–1928) deriving the formula for the force on a current-carrying conductor in a magnetic field, as illustrated in Figure 3.1.

$$F = BIl \tag{3.1}$$

where B is the magnetic field strength, l is the length of the current-carrying conductor, I is the current magnitude, and F is the Lorentz force.

If a flat coil of wire with radius r is placed in a magnetic field and mounted on an axis, as shown in Figure 3.2, then current flowing through the coil, with the polarities shown, will result in Lorentz forces developing on the coil wires, leading to creation of torque and rotation of the coil, where:

$$T = BIlr \sin \theta \tag{3.2}$$

As the coil rotates counter clockwise (CCW), the torque will vary in a sinusoidal fashion from a maximum ($\theta = 90°$) to zero ($\theta = 0°$) at which point motion would stop. However, if a second coil is mounted at an angle to the first (dotted in Figure 3.2), then it would create additional torque to maintain motion and rotate the first coil such that θ will increase beyond 0 and would again develop torque *if its current were reversed* in order to maintain CCW rotation.

The reversal of the current in the rotating coils is defined as *commutation*.

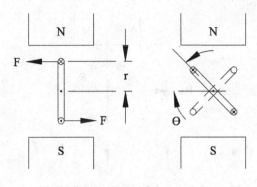

$$T \propto B \, l \, Ir\sin \Theta$$

Figure 3.2 Torque produced by coil in magnetic field

3.1.2 The Brush Motor

A typical motor will have numerous coils mounted on the rotor (the armature) connected to a series of copper segments, the commutator, mounted on one end of the rotor. Current flows through the coils via these segments that in turn are contacted by stationary soft carbon devices (the *brushes*). As the rotor rotates, the current in the coils will reverse direction in a cyclical fashion, maintaining torque and, therefore, rotation.

Although the coils rotate within a stationary magnetic field assembly a reciprocal arrangement could be constructed in which the coils are stationary and the magnet structure rotates. This is the basic concept of the brushless motor (Section 3.1.4) and the stepper motor (Section 3.1.9).

A schematic of a 12 coil, 12 segment commutator is shown in Figure 3.3.

Essentially, the action of commutation is a form of inverter, converting direct current into the motor into alternating current in the rotor coils. Strictly speaking, from the viewpoint of the internal functionality of the motor, there is no such device as a DC motor.

An alternate way of describing commutation is shown in Figure 3.4 which shows the permanent field B_F and the total field produced by the rotor B_A at an ideal 90° angle, which will result in maximum torque. The purpose of commutation is to maintain this 90° relation as much as possible.

As the rotor rotates, the total torque is the sum of the torques developed by the individual coils. Figure 3.5 shows the commutated waveforms and the total torque for the 12 coil, 12 segment rotor.

Figure 3.3 Twelve coil brush motor schematic

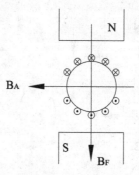

Figure 3.4   Ideal field orientation in DC motor

Note that the summation process results in the torque having a cyclical variation, called torque ripple. In this example the torque has a $\pm2\%$ nominal to peak variation.

The torque described by Equation 3.2 assumes that the current is constant and therefore sinusoidal torque components will be produced by each coil. This is only true if the BEMF waveshape (see the following) is a perfect square wave. At best it is trapezoidal and will result in torque ripple components in addition to those shown in Figure 3.5.

Torque ripple can be reduced by increasing the number of rotor coils and/or creating parallel coil paths. However, for a given overall diameter, this would require a larger quantity of smaller commutator segments, limited by practical mechanical constraints.

Starting in the mid 1800s literally thousands of rotor coil configurations, known basically as either lap or wave designs, with various winding pitches and parallel or series conductor paths, have been created to minimize torque ripple and maximize motor efficiency.

Originally the field assembly consisted of coils on salient poles powered with DC and wound to create alternate north and south poles.

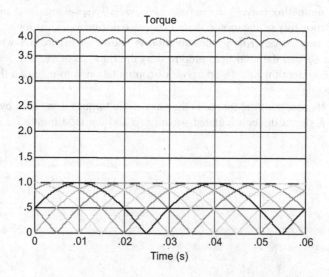

Figure 3.5   Commutated waveforms and total torque in 12 coil motor

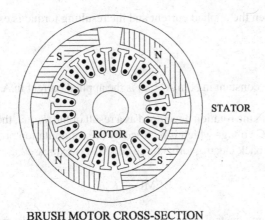

BRUSH MOTOR CROSS-SECTION

Figure 3.6 Permanent magnet brush motor cross-section

Beginning in 1930, a powerful permanent magnet material, Alnico (aluminum, nickel, cobalt) was created that has, to a large extent, replaced the wound stator poles and resulted in the development of today's permanent magnet brush motor.

In the 1960s a relatively inexpensive permanent magnet material, Hard Ferrite, a compound of various ceramics and iron oxide was developed, helping to create a class of low cost brush motors.

A cross-section of a typical four pole permanent magnet brush motor is shown in Figure 3.6.

3.1.2.1 Armature Control – Wound Field or Permanent Magnet Field

This is illustrated in Figure 3.7. Regardless of whether the field is created by current-carrying coils mounted on salient poles or by permanent magnets, armature control assumes that the field is constant and all control is dependent on control of the armature current.

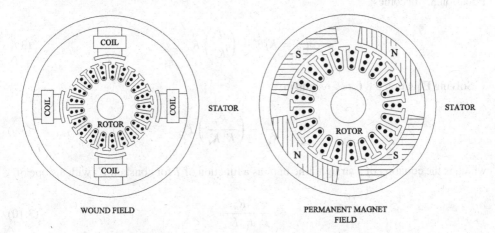

Figure 3.7 Wound field versus permanent magnet motor

The relation between the applied current and the resulting torque is expressed as:

$$T = K_t I_a \tag{3.3}$$

where K_t is the torque constant in N m A^{-1}, I_a is the applied current in A and T is the resulting torque in N m.

Simultaneously with the rotation developed as a result of the torque, the rotor coils will have induced in them an AC voltage which is rectified by the commutation process and opposes the applied voltage. This back electromotive force (BEMF) is expressed as:

$$\text{BEMF} = K_e \theta' \tag{3.4}$$

where K_e is the voltage constant in V rad^{-1} s^{-1}, θ' is the velocity in rad s^{-1} and the BEMF is in V.

In these units, K_t and K_e are numerically equal. (See also Section 2.6)

The rotor coils have both resistance (R_a) and inductance (L_a), resulting in the following equations relating the developed torque (T) as a result of creating the current (I) by applying the voltage (E_a), (Equations 2.38 and 2.40 repeated here):

$$E_a = K_e \theta' + I_a R_a + L_a I_a' \tag{3.5}$$

$$T = K_t I_a \tag{3.6}$$

Combining Equations 3.5 and 3.6 results in:

$$E_a = K_e \theta' + \left(\frac{T}{K_t}\right) R_a + L_a \left(\frac{T'}{K_t}\right) \tag{3.7}$$

Since in the steady-state T is constant and $T' = 0$, that is, no acceleration or deceleration, Equation 3.7 becomes:

$$E_a = K_e \theta' + \left(\frac{T}{K_t}\right) R_a \tag{3.8}$$

Solving Equation 3.8 for θ' results in:

$$\theta' = \frac{E_a}{K_e} - \left(\frac{R_a}{K_e K_t}\right) T \tag{3.9}$$

which is the equation of a straight line of θ' as a function of T for constant E_a with a slope of

$$-\frac{R_a}{K_e K_t} \tag{3.10}$$

This is the *speed/torque curve* of the motor.

If $T = 0$, then $\theta' = \dfrac{E_a}{K_e}$ which is the no load velocity   (3.11)

If $\theta' = 0$, then $T = \left(\dfrac{E_a}{R_a}\right) K_t$ which is the stall torque; $\dfrac{E_a}{R_a}$ being the stall current   (3.12)

As shown in Section 2.6, Equations 2.38 (Equation 3.5) and 2.40 (Equation 3.6) can be combined with Equation 2.39 (which describes the mechanical dynamics of the motor) resulting in the general velocity control simulation shown in Figure 2.22 with the characteristic Equation 2.42

$$s^2 + \frac{(JR_a + BL_a)}{JL_a}s + \frac{(K_eK_t + BR_a)}{JL_a}$$ (3.13)

If the third term of Equation 3.13 is factored as follows:

$$\frac{R_a\left(\dfrac{K_eK_t}{R_a} + B\right)}{JL_a}$$ (3.14)

the viscous damping term (B) has added to it the term $\frac{K_eK_t}{R_a}$ which is the reciprocal of the slope of the speed/torque curve (Equation 3.10).

Thus when performing initial design of a system and comparing various motors as candidates, comparing the slope of their speed/torque curves can be used as an indication of their contribution to the overall system damping.

The simulation shown in Section 2.6 is typically used to determine the time response of the motor (system) but it can also be used to display the speed/torque curve as shown in the following example.

Motor Data: $R_a = 2\ \Omega$ $K_t = 4032$ g cm A^{-1} $J = 8$ g cm s^2
$L_a = 20$ mH $K_e = 0.40$ V rad^{-1} s^{-1} $B = 14.4$ g cm rad^{-1} s^{-1}

The simulation of this motor with its response to an applied voltage of 43 V to achieve a steady state, no load velocity of 105 rad s^{-1} (1000 rpm) from zero to 1 s, followed by the application of a ramped load torque until stall is reached is shown in Figure 3.8.

Note that the load torque is simulated as a ramp rising from zero at $t = 1$ s to the stall value of 86 600 g cm at $t = 5$ s. (The load torque is multiplied by 0.001 at the plot input in order to display both the velocity and torque on the vertical axis scale.

The simulation plot can be converted to display speed versus load torque, that is, the speed/torque curve as shown in Figure 3.9, for both 43 V (105 rad s^{-1} no load speed) and 86 V (210 rad s^{-1} no load speed) input voltages.

See Section 3.1.5 for a more detailed review of speed/torque curves.

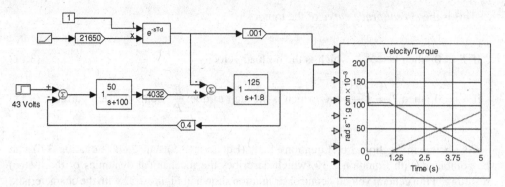

Figure 3.8   Step response of motor with ramped load to stall

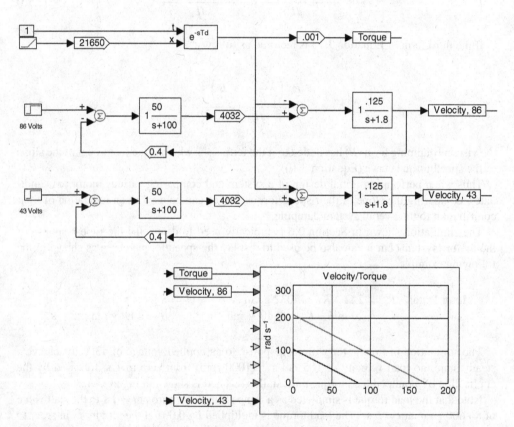

Figure 3.9   Speed/torque curves for simulated motor (available in full color at www.wiley.com/go/moritz)

3.1.2.2 Field Control – Wound Field

The field windings in a typical wound field motor usually have a comparatively large number of turns of fine wire to minimize the power needed to create the field. It would seem, therefore, that the motor could be controlled with less power than used in the armature control approach. Although this is true, there are negative aspects to field control.

The field windings have considerably higher inductance than the armature circuit, resulting in a lower frequency break point and larger response delays.

The magnetization curve of the field iron is nonlinear and becomes constant as saturation is approached. Therefore the field control current must be limited to a maximum value less than the saturation level in order for approximate linear control to be established.

In armature control it is assumed that the field is constant. Reciprocally, in field control, the armature must have constant excitation and, therefore, it must be supplied with a constant current.

In small motors, this can be accomplished approximately by having a large resistance in series with the armature so that the BEMF created in the armature has a limited effect on the armature current. This resistor will have large dissipation contributing to system inefficiency.

For large motors, a true constant current power supply is needed. This supply involves the use of power semiconductors and a feedback circuit topology almost as complex as the circuit used to control the armature in the armature control approach.

Assuming that the foregoing limitations are acceptable, a linear analysis can be developed to create the transfer function for the field control approach and then compare it to armature control.

For field control, Equations 2.38–2.40 can be written as:

$$E_f = I_f R_f + L_f I_f' \qquad (3.15)$$

$$T = J\theta'' + B\theta' + T_l \qquad (3.16)$$

$$T = K_f I_f \qquad (3.17)$$

Note that Equation 3.15 does not contain a velocity-dependent (BEMF) term.
Combining these equations will result in the following characteristic equation:

$$s^2 + \frac{JR_f + BL_f}{JL_f}s + \frac{R_f}{JL_f}B \qquad (3.18)$$

Compare this to Equations 3.13 and 3.14 and note that here the motor electrodynamic characteristics do not contribute to the overall damping, which then is determined purely by the mechanical effects of the motor and load.

3.1.2.3 Series Motor

A series motor is a wound field motor in which the field winding and the armature winding are connected in series and, therefore, have the same current.

A schematic for the series motor is shown in Figure 3.10a.

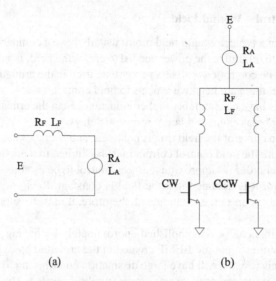

(a)   (b)

Figure 3.10   (a) Series motor schematic; (b) Split field series motor

Since the same current flows through the field and armature windings, reversing the polarity of the applied voltage will reverse the magnetic polarity of both the field and armature and, therefore, will not reverse rotation. In order to reverse rotation, the connection between the field winding and the armature has to be reversed, which can be accomplished with a DPDT power relay or by making the connection between the field and armature via an "H" bridge power stage. (see Section 3.1.3). A simpler way of achieving bi-directional control is to fabricate the motor with two field windings with opposite magnetic polarities and selecting the appropriate one for CW or CCW operation as shown in Figure 3.10b.

The developed torque (T) is proportional to the product of the armature and stator fields. The armature field strength is proportional to the current (I). The stator field strength is a nonlinear function of the current (I), as determined by the B/H curve of the stator iron.

Figure 3.11 shows a typical B/H curve (actual) together with a normalized linear approximation of the B/H curve, used in the following simulation.

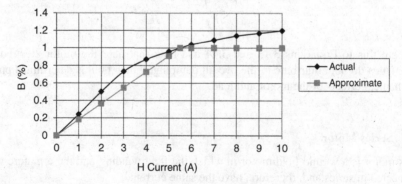

Figure 3.11   Stator field *B/H* curve; actual versus linear approximation

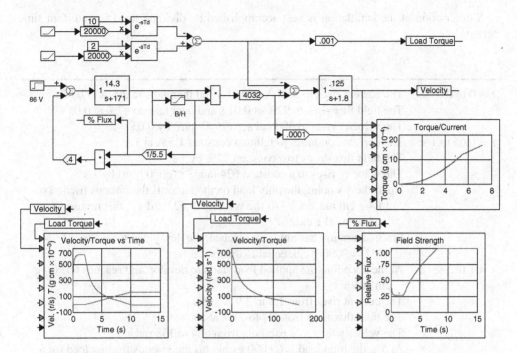

Figure 3.12   Series motor simulation (available in full color at www.wiley.com/go/moritz)

The result is that up to saturation of the stator iron, the torque is proportional to the product of the two field strengths. Beyond saturation, since the field strength is now constant, the torque is proportional only to the armature field strength [1].

Therefore, as load torque is increased up to the point of saturation the speed/torque curve is a parabola. Once saturation is reached, the speed/torque curve becomes linear.

Also, the BEMF is proportional to the product of the velocity and the strength of the stator field, again as determined by the *B/H* curve.

Figure 3.12 shows a simulation of a series motor, operating from 0 to 15 s, in response to a step application of 86 V with no load from 0 to 2 s, followed by a linearly increasing load up to stall at 10 s.

The motor is the same as that shown in Figures 3.8 and 3.9, but rewired for series operation with:

$$R_f = 10 \ \Omega \qquad L_f = 50 \ \text{mH}$$

resulting in:

$$R_{\text{total}} = 12 \ \Omega \quad L_{\text{total}} = 70 \ \text{mH}$$

(The 0.001 and 0.0001 factors have been used to allow the two inputs to the respective plots to use compatible vertical axis calibration).

A description of the simulation is best accomplished by dividing the 15 s into four time periods:

$t = 0$ to 0.05 s: The current rises to 5.05 A at 0.01 s and then falls to 2.6 A at 0.05 s

The field flux rises to 92% at 0.01 s and then falls to 47% at 0.05 s

The velocity rises to 300 rad s^{-1} (2900 rpm) by 0.05 s

$t = 0.05$ to 1 s: The current continues to fall to a constant 1.5 A at 1 s

The field flux drops to a constant 27% by 1 s

The velocity rises to a constant 624 rad s^{-1} (6000 rpm) by 1 s

Up to the 1 s point, the only load on the motor is the viscous friction of 14.4 g cm rad^{-1} s^{-1}. At the velocity of 624 rad s^{-1} this represents a load of 9000 g cm.

Since saturation has not been reached, the developed torque is: $1.5^2 \times 4032 = 9000$ g cm, equal to the load

$t = 1$ to 5 s: At the 2 s point, the applied load starts to develop and reaches 60 000 g cm at 5 s

The current rises from 1.5 to 3.9 A at 5 s

The field flux rises from 27 to 71% at 5 s

The velocity falls as a parabola from 624 to 138 rad s^{-1}

At 5 s, the total load is 60 000 g cm plus the viscous friction load for a total of 61 000 g cm.

Saturation has not yet been reached so the developed torque is $3.9^2 \times 4032 = 61\,000$ g cm, equal to the load

$t = 5$ to 15 s: At 8 s, the current reaches 5.5 A, at which point the field is saturated. From this point to the 10 s point, the field equivalent value remains at 5.5 A, the field flux is 100% and the velocity fall becomes linear.

At 10 s, stall is reached. The current rises linearly to 7.2 A at 10 s. The total load is 160 000 g cm of applied load since, at zero velocity, the load created by viscous friction is zero.

The developed torque is: $7.2 \times 5.5 \times 4032 = 160\,000$ g cm, equal to the load

---

Notice how at 5.5 A, (8 seconds) the torque versus current, the velocity versus time and the velocity versus torque curves all change from parabolas to straight lines.

3.1.3   The "H" Drive PWM Amplifier

Although there are many different amplifier designs at different bus voltage levels (24 to 600 V DC), current levels (5 to 50 A) and many different input and internal circuit topologies, they all share a common power output configuration: the "H" Design.

Figure 3.13 shows a general schematic for the "H" stage. It consists of four power semiconductors, four power rectifiers and a current sensing resistor.

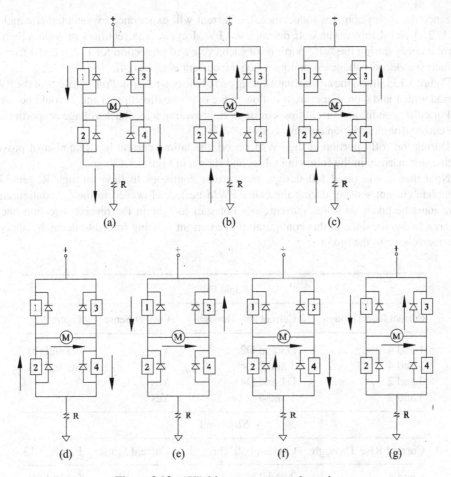

Figure 3.13 "H" drive power stage schematic

The power semiconductors (1, 2, 3 and 4) are either transistors or MOSFETS providing bi-polar application of voltage and current to the motor load. The rectifiers are either discrete devices or contained within the power devices.

The amplifier typically operates in PWM (pulse width modulation) mode at switching frequencies of 20 kHz (50 μs period) to 100 kHz (10 μs period).

Amplifiers vary in their design with respect to the duty cycle of the PWM period. Some allow the duty cycle to reach a maximum of 100% to achieve peak current as quickly as possible, while others force the duty cycle to be no larger than 95% to obtain more reliable control over possible over current conditions.

This plus the electrical time constant, the PWM frequency and the bus voltage must all be considered in determining the suitability of a particular amplifier to meet system requirements.

In general, the larger the electrical time constant the lower the PWM frequency should be, with the bus voltage determined by the K_e of the motor and the peak system velocity In addition, for reliable control of the PWM logic, it should take two or more duty cycles to reach peak current.

Since the motor contains inductance, the current will experience exponential rise and fall as 1, 2, 3 and 4 turn on and off during each PWM cycle. The rectifiers provide a path for current decay during the "off" portions of each cycle and protection for 1, 2, 3 and 4 from the inductive induced voltage excursions when the current is turned off.

Figure 3.13a shows the current path during one of the two possible "on" portions of the PWM period with 1 and 4 on. For current to flow in the opposite direction, 3 and 2 would be "on".

For either condition, current flows through R, providing a feedback voltage proportional to the current for closed loop current control.

During the "off" portion of the PWM period, the falling current has a number of possible paths, summarized in the following tables and shown in Figure 3.13b–g.

Note that in the rapid fall design, the current continues to flow through R, providing complete current sensing during the entire PWM period. However, in the (c) configuration, care must be taken such that current does not start to flow in the reverse direction once it has reached zero, since in this configuration the current is being forced to decay by applying reverse voltage to the motor.

Rapid Fall			
Current Rise Through	Current Fall Through	Current Sense	Figure 3.13
1 and 4	D3 and D2	Yes	(a) and (b)
1 and 4	3 and 2	Yes	(a) and (c)
3 and 2	D1 and D4	Yes	
3 and 2	1 and 4	Yes	

Slow Fall			
Current Rise Through	Current Fall Through	Current Sense	Figure 3.13
1 and 4	2 and 4	No	(a) and (d)
3 and 2	4 and 2	No	
1 and 4	1 and 3	No	(a) and (e)
3 and 2	3 and 1	No	
1 and 4	2 and D4	No	(a) and (f)
3 and 2	D4 and 2	No	
1 and 4	1 and D3	No	(a) and (g)
3 and 2	D1 and 3	No	

Additional "H" information is described in Section 3.1.10.

3.1.4   The Brushless Motor [2, 3]

In 1955 and 1956 two US patents (2719944 and 2753501) were issued to H.D. Brailsford in which are described motors commutated by transistors. These were not completely brushless in that in one version it required an external disturbance to begin operation and in the second version the motor had small brushes to initiate rotation which then disconnected from the commutator assembly by centrifugal force action as the motor accelerated to operating speed.

In October 7, 1962, T.C. Wilson and P.H. Trickey published an AIEE paper titled "DC Machine with Solid State Commutation", describing a true self-starting, brushless DC motor.

During the US space program of the 1960s and 1970s it was found that brush motors experienced short term brush life when operated in a high altitude (vacuum) environment. This led to the development of brushless motors to avoid this problem, which in turn has led to their commercial use in a wide range of applications.

This commercialization of brushless motors has been advanced primarily by the development of:

- High energy rare earth permanent magnets
- Low cost power semiconductors
- Hall device shaft position sensors
- Microprocessors with associated control software.

Advantages that brushless motors have over brush motors include:

- No EMI; no mechanical brush/commutator interface
- Higher speeds at high torque; higher power for smaller comparative size
- Lower thermal resistance
- Longer life; bearing rather than brush-dependent
- Better suited for clean room/explosion proof environment.

3.1.4.1   Construction

Figure 3.14 shows a cross-section of a typical brushless motor.
It consists of three main parts:

- The stator, wound with coils which are typically connected in a "wye" or "delta" configuration, similar to the stator in an induction or synchronous motor.
- The rotor, consisting of a soft iron core on which are bonded the permanent magnets creating the alternate north and south poles.
- A shaft position sensor (not shown) which can be:

   An assembly of three Hall sensors mounted on the housing plus a magnetized disc, having the same pole count as the rotor, mounted on the shaft.

   A digital encoder with a commutation track supplying three position signals.

   A resolver whose R to D converter supplies three commutation signals.

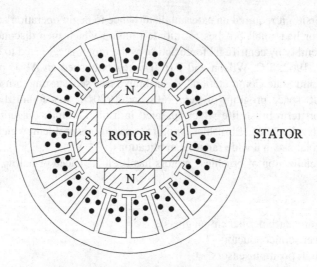

Figure 3.14   Brushless motor cross-section

3.1.4.2   Winding Configuration

A schematic for an exact brushless duplicate of the 12 winding brush motor coil connection shown in Figure 3.3 is shown in Figure 3.15a.

It consists of 12 stator coils, each separately driven by a bi-polar H bridge amplifier requiring 24 wires between the motor and amplifiers. In addition it requires a shaft position sensor supplying decodable information to reverse the current flow in pairs of coils every 30° of shaft rotation for a two pole version.

Such an assembly for both the motor and the drive electronics would be highly impractical, both economically and technically.

By re-configuring the windings into a three phase design with appropriate electromagnetic pole distribution, the windings can be connected either as a "wye" or "delta" and driven by three half bridges as shown in Figure 3.15b which has become the industry standard stator design for brushless motors.

3.1.4.3   Trapezoidal BEMF/Six Step Drive

The winding can be designed such that the BEMF per phase is trapezoidal, as in Figure 3.16.

In addition to the BEMF, Figure 3.16 shows the currents that are made to flow in each phase, during the portion of each cycle when the BEMF is a constant value.

The drive electronics is arranged such that current is made to flow through two of the three phases for each 60° portion of the cycle. In each 60° portion, current flows into one phase and out of the connecting phase, resulting in six current conditions for a complete 360° cycle. Figure 3.17 shows the simple decoding of the three encoder signals to produce the six drive signals.

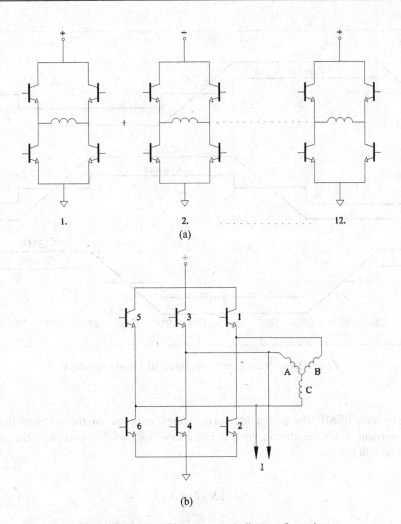

Figure 3.15 Brushless motor winding configuration

Refer to Figures 3.16, 3.17 and 3.15(b) to follow the current flow for a complete commutation cycle as follows:

Period	Phases	Drive Signal/Transistor
30–90	$A + \bar{B}$	$3 + 2$
90–150	$A + \bar{C}$	$3 + 6$
150–210	$B + \bar{C}$	$1 + 6$
210–270	$B + \bar{A}$	$1 + 4$
270–330	$C + \bar{A}$	$5 + 4$
330–390	$C + \bar{B}$	$5 + 2$

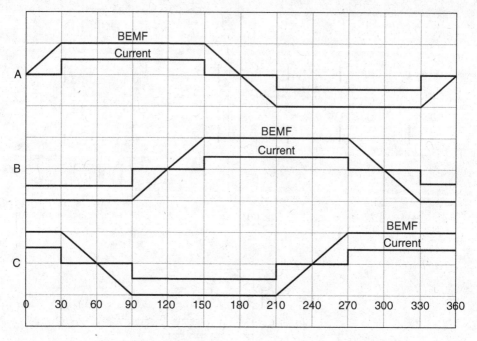

Figure 3.16   Waveforms; trapezoidal BEMF/six step drive

Since both the BEMF (the fixed field) and the current are constant, the torque produced will also be constant. Since the current flows in two phases for each 60° portion of the cycle, the total torque will be:

$$T = 2K_X I = K_T T \tag{3.19}$$

as in a brush motor. Since the BEMF and the current are constant, the torque will have no cyclical variation and, therefore, will theoretically have no cogging (see Section 3.1.4.6).

A simulation of this is shown in Figure 3.18, in which 1 A DC three phase currents (see Figure 3.17) are applied to 1 V phase to neutral three phase trapezoidal BEMF waveforms. For each phase the two signals are multiplied and then the three results are summed to obtain the total constant torque.

3.1.4.4   Sine BEMF/Sine Drive [4]

The winding can also be designed to have a sinusoidal, three phase BEMF and be driven by three phase sine wave power. In this case, all three phases are in operation simultaneously in a 1, 120, 240 phase relationship and the total torque will be the arithmetic sum of the three phase torques.

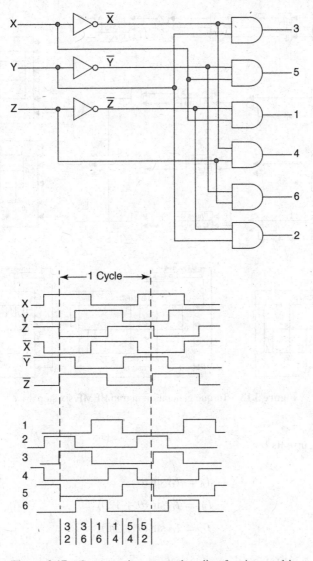

Figure 3.17 Commutation sensor decoding for six step drive

Let the three BEMFs be:

$$B_A = B_O \sin \theta \qquad (3.20)$$

$$B_B = B_O \sin (\theta + 120) \qquad (3.21)$$

$$B_C = B_O \sin (\theta + 240) \qquad (3.22)$$

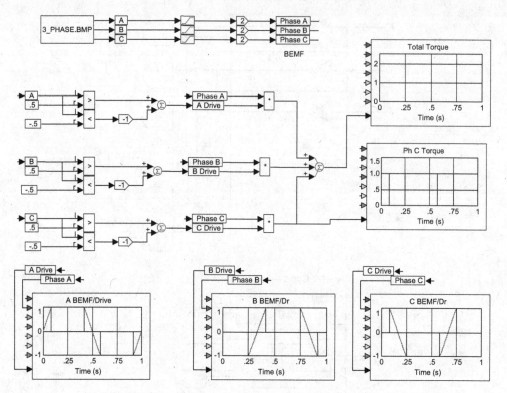

Figure 3.18   Torque generation in trap BEMF/six step drive

Let the three currents be:

$$I_A = I_O \sin \theta \tag{3.23}$$

$$I_B = I_O \sin (\theta + 120) \tag{3.24}$$

$$I_C = I_O \sin (\theta + 240) \tag{3.25}$$

The total torque will then be the sum of the three separate phase torques:

$$T = T_A + T_B + T_C = I_A B_A + I_B B_B + I_C B_C \tag{3.26}$$

$$= I_O B_O \left(\sin^2 \theta + \sin^2 (\theta + 120) + \sin^2 (\theta + 240) \right) = \frac{3}{2} I_O B_O = K_T I_O \tag{3.27}$$

showing that the theoretical result is a constant, cogging-free torque as a function of the applied current (see Section 3.1.4.7) [5, 6].

A simulation of this is shown in Figure 3.19, in which 1 Apeak three phase currents are applied to 1 V phase to neutral three phase BEMF waveforms. For each phase the two signals are multiplied and then the three results are summed to obtain the total torque.

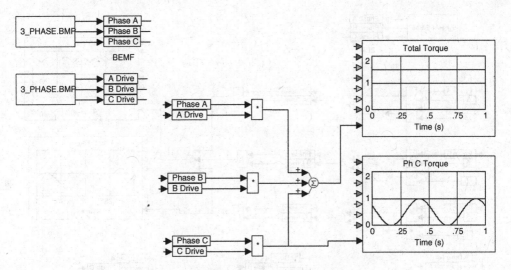

Figure 3.19   Torque generation in sine BEMF/sine drive

Since both the voltage and torque constant are required to determine the voltage and current rating of an amplifier to be used to drive a particular motor, it is necessary to measure these parameters. Measuring the voltage constant is straightforward by driving the motor with a second motor at 1000 rpm and measuring the peak BEMF, providing K_E in V/krpm. Measuring the torque constant is not as simple, but it can be derived from the voltage constant as follows:

Output power $= 0.105\,NT$ W, with N in rpm and T in N m.

$$0.105NT = \sqrt{3}\,E_{RMS}I_{RMS} \quad \text{(three phase power)} \tag{3.28}$$

$$I_{RMS} = \frac{0.105NT}{\sqrt{3}\,E_{RMS}} = \frac{(0.105)\,NT}{\dfrac{\sqrt{3}\,E_{PEAK}}{\sqrt{2}}} = \frac{105T}{K_E}\sqrt{\frac{2}{3}} \tag{3.29}$$

$$K_{T/RMS} = \frac{T}{I_{RMS}} = \frac{K_E}{105}\sqrt{\frac{3}{2}} = 0.0117K_E\ \frac{\text{N m}}{\text{A}} \qquad K_{T\,PEAK} = 0.0083K_E\ \frac{\text{N m}}{\text{A}} \tag{3.30}$$

where K_E is the peak voltage constant in V/krpm.

Note: This is different from the case of the brush motor, where:

$$K_T = 0.00955K_E\ \frac{\text{N m}}{\text{A}}$$

where K_E is the voltage constant in V/krpm.

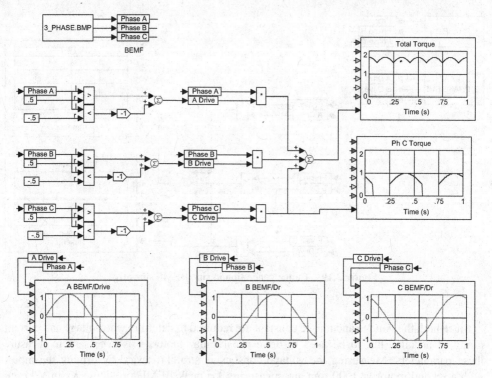

Figure 3.20   Torque generation in sine BEMF/six step drive

3.1.4.5   Sine BEMF/Six Step Drive

Since the six step brushless amplifier is the most economical, it is of interest to determine the result when it is used to drive a motor with sinusoidal BEMF, since it is becoming the pseudo-standard brushless motor. Figure 3.20 shows a simulation of this combination.

The drive currents are the same as shown in Figure 3.18 for the six step drive but acting on sine BEMF. The result is an average of 1.62 compared to 1.5, or 8% higher torque, but with a ripple content of ±7%.

Since the torques are less than 10% different, the decision on which drive to use would be based on the effect that the ripple will have on system performance. In a velocity control system with sufficient bandwidth and inertia the ripple would have minimal effect, but a simulation should be performed to assure this.

In a positioning system, where reliable torque control is required around zero velocity it is best to use the sine/sine approach, even though it requires the increased expense of an encoder or resolver plus the generation of the sine drive in the amplifier.

3.1.4.6   Cogging/Ripple Torque

Cogging and ripple effects tend to be thought of as one and the same effect, namely torque variations causing cyclical acceleration/deceleration within each commutation cycle per revolution. Both have their most negative torque variation effect at low speed and when settling into final position.

However, they are different:

- **Cogging** is a function of the motor structure and is independent of the drive electronics. It is a torque perturbation caused by the attraction of the rotor magnets to the stator iron to minimize the reluctance. It is a function of the number of rotor poles and stator poles and the ratio of the two quantities. The usual method to minimize cogging is to skew the stator slots. Cogging can be felt by simply rotating the motor shaft of an unpowered motor, and feeling the detent action. A well designed motor will have low cogging, less than 1% of rated torque, which can be filtered by the action of the system inertia and may be less than the ripple torque.

    Since it is a negative characteristic of the motor, it is never specified in typical motor data sheets and must be determined by contacting the manufacturers' application engineers.

- **Ripple torque** is generated as a result of less than perfect wave shapes and interaction between the motor and the drive electronics.

    It is also known as commutation torque ripple and the ripple frequency is proportional to motor speed and pole count. In general, the lower the speed of the system the higher should be the pole count in order to enable filtering of the ripple by loop compensation (PI, etc.).

    Ripple torque can be caused by one or more of the following:
    - BEMF amplitude in one phase is different from the other phases
    - Current amplitude in one phase is different from the other phases
    - BEMF waveshape in either trapezoid or sinusoidal is not ideal
    - Current waveshape in either trapezoid or sinusoidal is not ideal
    - Phasing between commutation sensing and BEMF is misaligned
    - Commutation sensors are not 120° separated
    - Wave shapes of BEMF or current contain torque-producing harmonics.

3.1.5   Speed/Torque Curves

Figures 3.21–3.23 are speed/torque plots for a brushless DC motor, with parameters as listed in the left column.

As described in Section 3.1.2, a speed/torque curve displays the variation in speed (as the dependent variable) as the load torque (the independent variable) is increased from zero to some maximum value, with constant applied voltage.

Rather than being a single curve, modern speed/torque plots also define speed/torque zones, delineating areas in which the motor can operate, limited by boundaries determined by voltage, temperature and current (torque) as follows:

- The upper, downward sloping curve is the actual speed/torque curve described in Section 3.1.2. It starts at the no load speed (3700 rpm for this example) and decreases as the load increases per Equation 3.9.
- The "continuous" zone is that in which the motor can operate at any combination of speed and torque without time limitations and without exceeding its maximum rated temperature.

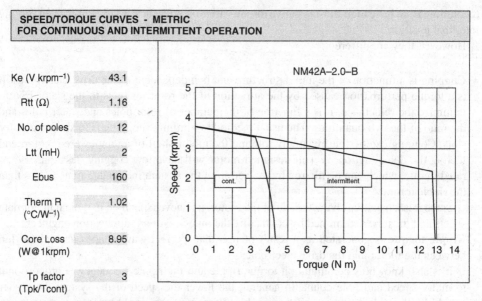

SPEED/TORQUE CURVES - METRIC
FOR CONTINUOUS AND INTERMITTENT OPERATION

Ke (V krpm⁻¹)	43.1
Rtt (Ω)	1.16
No. of poles	12
Ltt (mH)	2
Ebus	160
Therm R (°C/W⁻¹)	1.02
Core Loss (W@1krpm)	8.95
Tp factor (Tpk/Tcont)	3

NOTE: MAXIMUM VOLT. LIMITED NO LOAD SPEED (krpm) = 3.71 Tamb = 25

Figure 3.21   Speed/torque curve, R @ 155 °C, amb. @ 25 °C

The vertical boundary separating the "continuous" zone from the "intermittent" zone is the locus of the maximum rated temperature, usually 130 or 155 °C, in a specified ambient temperature, usually 25 or 45 °C.

For example, in Figure 3.21, for two points along the boundary:
(a) at stall, velocity = 0, $T = 4.3$ N m

$$I = \frac{4.33}{0.412} = 10.5 \text{ A}$$

$$\text{copper loss} = I^2 R_{155} = 10.5^2 \,(1.16) = 126 \text{ W}$$

$$\text{velocity loss} = 0$$

$$\text{temperature} = 126 \times 1.02 + 25 = 155 \,^\circ\text{C}$$

(b) at velocity = 1600 rpm, $T = 4$ N m.

$$I = \frac{4}{0.412} = 9.72 \text{ A}$$

$$\text{copper loss} = I^2 R_{155} = 9.72^2 \,(1.16) = 110 \text{ W}$$

$$\text{velocity loss} = (1.6)^{1.5} \,(8.95) = 18 \text{ W}$$

$$\text{temperature} = (110 + 18) \times 1.02 + 25 = 155 \,^\circ\text{C}$$

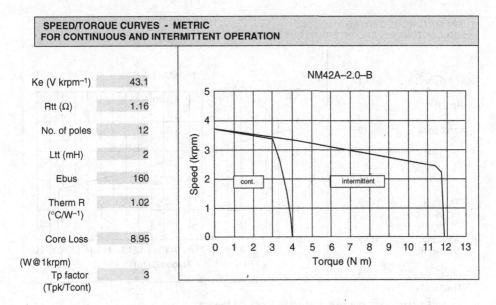

SPEED/TORQUE CURVES - METRIC		
FOR CONTINUOUS AND INTERMITTENT OPERATION		
Ke (V krpm⁻¹)	43.1	
Rtt (Ω)	1.16	
No. of poles	12	
Ltt (mH)	2	
Ebus	160	
Therm R (°C/W⁻¹)	1.02	
Core Loss	8.95	
(W@1krpm)		
Tp factor (Tpk/Tcont)	3	

NOTE: MAXIMUM VOLT. LIMITED NO LOAD SPEED (krpm) = 3.71 Tamb = 45

Figure 3.22   Same as Figure 3.21, with amb @ 45 °C

Note that the curvature of the boundary is due to the velocity dependent losses (core and viscous damping). If these losses are negligible (or ignored) this boundary would be a straight vertical line.

- The right vertical boundary of the "intermittent" zone is determined by the maximum peak current allowable without causing permanent demagnetization of the rotor magnets. It is usually two to ten times the maximum continuous stall torque. In this example it is three times.

The length of time operation can remain in the "intermittent" zone depends on the magnitude of torque, the thermal resistance and the thermal time constant. Unfortunately, the thermal time constant is rarely available in product literature or even by direct contact with motor manufacturers. It must be measured as in Section 3.1.6.3 and applied as shown in Section 3.1.6.7.

Figure 3.21 shows the speed/torque data with resistance at its maximum 155 °C value in a 25 °C ambient, with the motor mounted on a 230 cm² aluminum plate.

Figure 3.22 is the same as Figure 3.21 except the ambient temperature is 45 °C.

Figure 3.23 is the same as Figure 3.21 except the motor is mounted on a 645 cm² aluminum plate.

3.1.6   Thermal Effects

There are a number of thermal effects, reviewed in the following, to be considered when determining the suitability of a particular motor for an application.

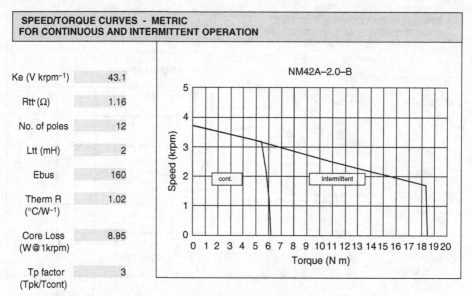

Figure 3.23   Same as Figure 3.21, with three times larger mounting plate

3.1.6.1   Heat Sources

- Copper losses: typically called I^2R losses, in which the current primarily supports the load, but also includes the current required to support the following losses.
- Coulomb friction losses: losses created by the friction in the motor bearings plus any additional friction in the load assembly. For example, the ways friction in a slide mechanism. Coulomb friction torque is independent of velocity.
- Viscous friction losses: losses in which the torque is proportional to the velocity (see Section 2.6).
- Core losses: losses created by hysteresis and eddy current effects in the steel laminations of the stator. Very rarely specified in motor data sheets and typically a minor effect at low to medium velocities (up to approximately 5000 rpm), but can be a considerable quantity in small motors operating at high speeds 10 000 to 30 000 rpm).

Example

A brushless motor with the following specifications is operating at 3000 rpm with a load torque of 3.3 N m.

$$K_t = 0.41 \text{ N m A}^{-1} \qquad K_e = 43.1 \text{ V/krpm} \qquad R_{155} = 1.16 \ \Omega$$

thermal resistance = 0.7 °C W^{-1}

bearing friction = 144 g cm

viscous friction constant, $B = 0.5$ g cm rad^{-1}s^{-1}

core losses = 8.95 W @ 1000 rpm

$$P_{out} = \frac{3.3 \times 3000}{7120} = 1.4\,HP = 1037\,W \qquad (3.31)$$

$$I_{load} = \frac{3.3}{0.41} = 8.05\,A \qquad (3.32)$$

$$BEMF = 43.1 \times 3 = 129.3\,V \qquad (3.33)$$

$$\text{Coulomb friction loss: } P_{coulomb} = 144\,g\,cm\ @\ 3000\,rpm = 4.4\,W \qquad (3.34)$$

$$I_{coulomb} = \frac{4.4}{129.3} = 34\,mA \qquad (3.35)$$

$$\text{viscous friction loss: velocity} = 3000\,rpm = 314\,rad\,s^{-1} \qquad (3.36)$$

$$T_{viscous} = 0.5\,g\,cm\,rad^{-1}\,s^{-1} \times 314\,rad\,s^{-1} = 157\,g\,cm \qquad (3.37)$$

$$P_{viscous} = \frac{157 \times 3000}{7.26 \times 10^7} \times 746 = 4.84\,W \qquad (3.38)$$

$$I_{viscous} = \frac{4.84}{129.3} = 37\,mA \qquad (3.39)$$

$$\text{core losses: } P_{core}\ @\ 3000\,rpm = 8.95 \times 3^{1.5} = 46.5\,W \qquad (3.40)$$

$$I_{core} = \frac{46.5}{129.3} = 0.36\,A \qquad (3.41)$$

$$I_{total} = I_{load} + I_{coulomb} + I_{viscous} + I_{core} = 8.5\,A \qquad (3.42)$$

$$\text{copper loss}\,(I^2 R) = 1.5\,(8.5^2 \times 1.16) = 126\,W \qquad (3.43)$$

$$\text{total dissipation} = I^2 R + P_{coulomb} + P_{viscous} + P_{core} = 182\,W \qquad (3.44)$$

$$t_{wdg} = 182 \times 0.7 + 25 = 155\,°C \qquad (3.45)$$

$$\text{efficiency, } \eta = \frac{1037}{1037 + 182} \times 100 = 85\% \qquad (3.46)$$

3.1.6.2 Thermal Paths

- Brush Motor: In the brush motor, the $I^2 R$ energy primarily flows from the armature coils, through the armature steel lamination stack to the shaft/bearing assembly and then to the attached load.

   Secondly, it conducts through the stator/rotor air gap, through the stator assembly (either permanent magnets or stator lamination/coil assembly), through the clearance gap between the stator OD and the housing ID, through the housing to the mounting surface which for test purposes is typically simulated by an aluminum plate.
- Brushless Motor: In the brushless motor, the $I^2 R$ energy is developed in the stator and therefore need only flow through the clearance gap between the stator OD and the housing ID, then through the housing to the mounting surface, again simulated by an aluminum plate.

   However, some energy will flow across the stator/rotor air gap and increase the temperature of the rotor magnets.

3.1.6.3 Thermal Resistance and Thermal Time Constant

- **Thermal resistance** is a measure of the temperature difference that exists between two thermal surfaces divided by the amount of heat energy flowing between them.

  The energy flow is always from the higher temperature to the lower temperature.

  The unit for thermal resistance is $°C\ W^{-1}$.

  Catalog values for thermal resistance are usually given as the value from winding to ambient as measured with the motor mounted on a particular size of aluminum plate. The size of the plate is not always given, which makes it difficult to coordinate the thermal resistance with the actual mounting condition in the application.

  Also, different manufacturers will test motors with identical motor constants on different size plates. A motor mounted on a 25×25 cm plate will have a larger stall torque rating than the same motor mounted on a 15×15 cm plate only due to the $3 : 1$ difference in the plate surface area.
- **Thermal time constant** is the time it takes for a thermal surface to reach 63% of its' final temperature value when undergoing an energy change. If the energy change is a step, then the temperature change will be exponential and for all practical purposes is usually considered to complete in four or five time constants.

  Thermal time constants are rarely, if ever, published in motor catalogues. Typical application notes concentrate on the RMS and peak torques being in the continuous region of the speed/torque curve to assure that the internal motor temperature does not exceed the maximum rating (either 155 or $130\,°C$), in which case the thermal time constant need not be considered, regardless of whether the operation is continuous or intermittent.

  In intermittent operation in which the torque exceeds the continuous rating into the intermittent region, the time constant must be considered in connection with the duty cycle to assure that the maximum rating is not exceeded, since remaining in the intermittent region for more than four or five thermal time constants will cause the temperature to exceed the maximum rating.

  Thermal time constants in brushless compared to brush motors can be decidedly different. In a brushless motor, due to the efficient thermal path between the stator and case, the stator-to-ambient time constant and the case-to-ambient time constant are essentially the same.

  However, in a brush motor, the relatively poor thermal path from the rotor to ambient and the relatively high thermal barrier from rotor to case results in a rotor-to-ambient time constant much lower than the case-to-ambient time constant.

  For example, a motor in which the rotor-to-ambient time constant is one minute has a case-to-ambient time constant of 15 minutes. Due to this large thermal lag, the case temperature cannot be used as a measure of the rotor temperature.
- Thermal resistance and time constant measurement.

  Both parameters can be measured by performing the following test:
  1. Mount the motor on a typical test plate or, if available, on the actual assembly on which it will be operating.
  2. Measure the winding resistance and record the ambient temperature.
  3. Supply current to the motor from a DC constant current power supply. For a brush motor, block the shaft from rotating.
  4. In addition, mount a thermocouple on the case.

5. At specific times, after power is applied, record the power supply voltage and the case temperature until three successive readings show no change.
6. Using the final voltage and current values, calculate the power supply output power and the hot resistance.
7. Use the hot resistance value to calculate the hot internal temperature.

$$t_{hot} = 254.5 \left(\frac{R_{hot}}{R_{20}} - 1 \right) + 20 \text{ for } t_{amb} = 20\,°C \qquad (3.47)$$

$$t_{hot} = 259.5 \left(\frac{R_{hot}}{R_{25}} - 1 \right) + 25 \text{ for } t_{amb} = 25\,°C \qquad (3.48)$$

(see Section 6.8)

8. Use the internal and ambient temperatures and the power to calculate the internal-to-ambient thermal resistance.
9. Use the case and ambient temperatures and the power to calculate the case-to- ambient thermal resistance.
10. Plot the voltage and thermocouple data points versus time, locate the points that are 63% of the final values of each to determine the thermal time constants of the winding and case, respectively.

3.1.6.4 Temperature Limits

- **Insulation**

The maximum allowable winding temperature is determined by the grade of insulation used for the copper coating and for slot lining. There are basically four grades of insulation, as shown below. If the winding reaches the maximum value, then the 25 or 20 °C winding resistance will increase by the multiplier shown in the third and fourth columns. (See also Section 6.8).

Grade	Max Temp	R_{max}/R_{25}	R_{max}/R_{20}
A	105	1.3	1.33
B	130	1.4	1.43
F	155	1.5	1.53
H	180	1.6	1.63

- **Permanent Magnets**

During manufacture the magnets are charged to saturation at the prevailing ambient temperature. Thereafter, they will be subject to elevated temperatures based on the system operating conditions.

Permanent magnet material can experience three basic demagnetization effects, known as reversible, irreversible or complete demagnetization.

Reversible demagnetization occurs in every permanent magnet based motor when operation is limited to the continuous/intermittent zones defined by the speed/torque curve and if the winding temperature does not exceed the maximum value for the particular insulation grade and the current does not exceed the maximum value per the data sheet. All the typically used magnet materials experience a decrease in their energy level as follows:

$$SmCo = -0.045\%/°C$$

$$NdFeB = -0.1\%/°C$$

$$Alnico = -0.01\%/°C$$

$$Ferrite = -0.2\%/°C$$

Although there is a thermal drop between the heat source (the winding) and the magnets, over the long term of operation, the magnet temperature can be assumed, as a worst case condition, to reach the same temperature as the winding, in which case the K_t and K_e will each decrease by the following factors, assuming the magnets were initially charged at 25 °C:

Temp	SmCo	NdFeB	Alnico	Ferrite
105	0.96	0.92	0.99	0.84
130	0.95	0.90	0.99	0.79
155	0.94	0.87	0.98	0.74
180	0.93	0.84	0.98	0.69

In general, when analyzing system performance using rare earth based motors, assume that K_t and K_e will be 12% lower than the 25 °C data sheet value.

Irreversible demagnetization, which depends on the specific design of a motor's magnetic circuit, will occur at temperatures considerably higher than those listed above. It will not be experienced if a motor is operated within the temperature and current limits specified on the data sheet. When returned to ambient the magnet will retain charge but at a level lower than its original saturation level.

Complete demagnetization will occur at temperatures known as the Curie temperature, at which point the magnet will lose all of its charge even after the temperature returns to ambient, that is, it will have no magnetism.

Curie Temp (°C)	
SmCo:	800
NdFeB:	320
Alnico:	850
Ferrite:	450

3.1.6.5 Actual Resistance Increase

In motor sizing, when calculating the copper losses to determine the elevated winding temperature, it is typical to use the maximum elevated resistance value, even though the final dissipation is less than the dissipation value that would produce the maximum resistance value. Essentially, this creates a "built-in" safety factor.

It is interesting to use a more exact calculation to determine the approximate value of this safety factor.

Assume that the non-copper losses (see Section 3.1.6.1) are negligible. Then the typical method to calculate the elevated temperature would be to use the following:

$$t_2 = \left(I_{RMS}^2 R_2 \right) (\theta) + t_0 \tag{3.49}$$

where t_2 is the final elevated temperature, I_{RMS} is the worst case RMS current, R_2 is the maximum resistance for the insulation grade being used, per Section 3.1.6.4, θ is the thermal resistance from winding to ambient, and t_o is the ambient temperature.

A more exact expression for the elevated temperature would be the following, in which R_2 is replaced by the expression for the increase in resistance of a copper winding as a function of temperature for an ambient temperature of 25 °C (See Section 6.8)

$$t_2 = I_{RMS}^2 \{ R_0 \left[1 + 0.00385 \left(t_2 - t_0 \right) \right] \} (\theta) + t_0 \tag{3.50}$$

This can be solved for t_2:

$$t_2 = \frac{I_{RMS}^2 R_0 \theta \left(1 - 0.00385 t_0 \right) + t_0}{1 - 0.00385 I_{RMS}^2 R_0 \theta} \tag{3.51}$$

Now, assume a system in which:

$$I_{RMS} = 7.3 \text{ A} \quad R_0 = 0.77 \; \Omega \quad \theta = 1.02 \; °C \, W^{-1} \quad t_0 = 25 \; °C$$
$$R_{155} = 1.16 \; \Omega$$

Then, per Equation 3.49,

$$t_2 = 7.3^2 \times 1.16 \times 1.02 + 25 = 88 \; °C \tag{3.52}$$

and, per Equation 3.50,

$$t_2 = \frac{7.3^2 \times 0.77 \times 1.02 \left(1 - 0.00385 \times 25 \right) + 25}{1 - 0.00385 \times 7.3^2 \times 0.77 \times 1.02} = 75 \; °C \tag{3.53}$$

with,

$$R_2 = 0.77 \left[1 + 0.00385 \left(75 - 25 \right) \right] = 0.92 \; \Omega \tag{3.54}$$

resulting in the actual elevated temperature being 15% lower than the value predicted by the approximate method and the elevated resistance value being only 20% rather than 50% higher.

3.1.6.6   Thermal Runaway

Examination of the denominator $(1 - 0.00385 I_{RMS}^2 R_0 \theta)$ of Equation 3.51 shows that if

$$0.00385 I_{RMS}^2 R_0 \theta = 1, \text{ then the denominator} = 0 \text{ and } t_2 \to \infty.$$

That is, if the energy being generated cannot be dissipated by the thermal sink to which it is connected and the temperature stabilizes at some elevated value, then the temperature will continue to rise until failure occurs – the motor will burn out!

Consider example (a) in Section 3.1.5 (reference Figure 3.21):

$$1 - (0.00385)\left(10.5^2\right)(0.77)(1.02) = 0.667$$

and the temperature will rise to a stable 155 °C.

However any significant increase in I_{RMS} or θ or the ambient temperature or a combination of all three could cause thermal runaway to occur.

The following Section 3.1.6.7, shows the development of a thermal model of the motor, including the effect of the thermal time constant showing the length of time the peak current can exist in the intermittent zone before the maximum rated temperature is reached.

3.1.6.7   Thermal Model

Figure 3.24 shows a simulation of Equation 3.50 with a single time constant transfer function added to represent the thermal time constant.

The values shown are for example (b) in Section 3.1.5. The thermal time constant shown was measured per Section 3.1.6.3 as 360 s, with the final temperature of 155 °C being reached in 1800 s (five time constants) and the resistance rising to 1.16 Ω.

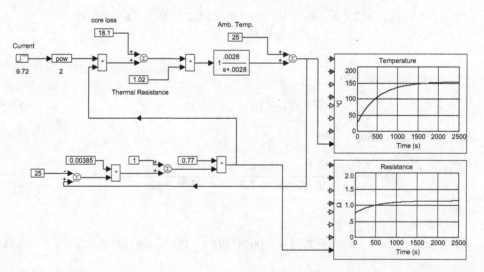

Figure 3.24   Thermal model with thermal time constant (available in full color at www.wiley.com/go/moritz)

By running this thermal model for various values of torque in the intermittent zone, the amount of time the torque (current) can exist before the temperature reaches the maximum rated value can be determined. The following table shows the result for the torque ranging, at zero velocity, from the maximum continuous value to the maximum peak value.

Torque (Nm)	Current (A)	Time to 155 °C (s)
4.33	10.5	1800 → ∞
5.0	12.14	632
6.0	14.56	329
7.0	17.0	213
8.0	19.42	152
9.0	21.84	116
10.0	24.27	91
11.0	26.7	74
12.0	29.13	61
13.0	31.55	51

This shows that if it is possible to measure the thermal resistance and the thermal time constant it is possible to determine, to a reasonable degree of accuracy, the length of time operation can exist in the intermittent zone.

In this example, the maximum peak torque can be obtained for almost 1 min before reaching the maximum temperature rating (155 °C).

However, if this were to occur on a repetitive basis, then these peak torques could not occur more frequently than every 30 min since it would take five thermal constants to cool down from 155 to 25 °C.

Fortunately, in the majority of motion control applications, peak torques (i.e., acceleration and deceleration torques) typically exist for only hundreds of milliseconds with the result that maximum temperature ratings are rarely reached.

Figure 3.25 is a typical torque timing diagram for a repetitive motion application.

In the following two examples the thermal simulation shows the temperature resulting for a condition first in which the RMS and the peak torques both fall in the continuous zone and second in which the RMS torque falls in the continuous zone and the peak torque falls in the intermittent zone.

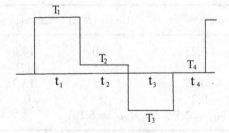

Figure 3.25 Repetitive motion timing diagram

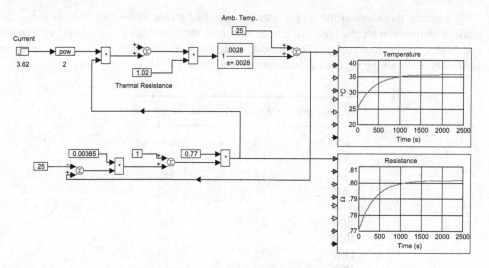

Figure 3.26   Example 1 temperature response

Example 1

Time (s)	Torque (N m)	Current (A)
$t_1 = 0.5$	3.5	8.5
$t_2 = 2.0$	0.5	1.2
$t_3 = 0.5$	3.0	7.3
$t_4 = 2.0$	0	0
$T_{RMS} = 1.5$ N m	$I_{RMS} = 3.62$ A	

Figure 3.26 shows the thermal simulation resulting in a final temperature of 36 °C and a final resistance of 0.8 Ω

Example 2

Time (s)	Torque (N m)	Current (A)
$t_1 = 0.25$	6.0	14.6
$t_2 = 2.0$	0.5	1.2
$t_3 = 0.25$	5.5	13.3
$t_4 = 0.2$	0	0
$T_{RMS} = 2.5$ N m	$I_{RMS} = 6.1$ A	

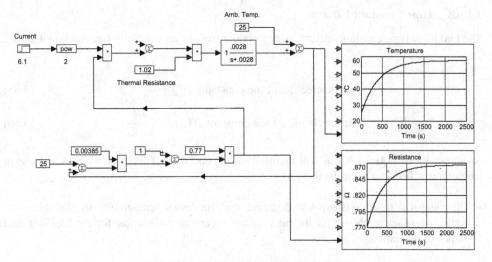

Figure 3.27   Example 2 temperature response

Figure 3.27 shows the thermal simulation resulting in a final temperature of 58 °C and a final resistance of 0.87 Ω.

Figures 3.26 and 3.27 show the final temperatures and resistances based on the RMS values.

Figure 3.28 shows a simulation of the actual 5 s current profile and the result of operating the profile repetitively for 1400 s, showing the "heating/cooling/heating/cooling, and so on." approaching the final RMS values.

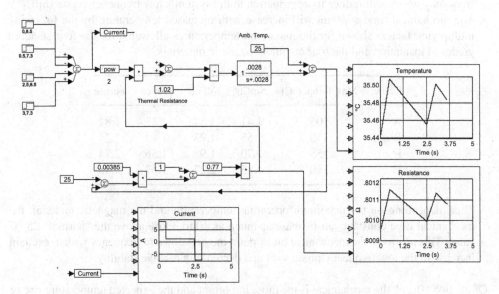

Figure 3.28   1400 s response of repetitive 5 s current profile

3.1.6.8   Time Constant Effects

The two basic time constants determining system response and stability (see Section 4.2) are:

$$\text{The electrical time constant: } TC_E = \frac{L}{R} \tag{3.55}$$

$$\text{The mechanical time constant: } TC_M = \frac{JR}{K_e K_t} \tag{3.56}$$

As shown in Section 3.1.6.4, R will increase and K_e and K_t will decrease with an increase in temperature. The result will be that:

- The electrical time constant will decrease with increased temperature by the following multiplying factors shown for the maximum temperatures allowable for the four standard grades of insulation.

Grade	Max. Temp. (°C)	TC_E Multiplier
A	105	0.77
B	130	0.71
F	155	0.67
H	180	0.63

A decrease in the electrical time constant corresponds to an increase in the electrical band width which in general is a positive result since the electrical break point will move up in frequency which will reduce its contribution to the system's low frequency phase shift.
- The mechanical time constant will increase with increased temperature by the following multiplying factors shown for the maximum temperatures allowable for the four standard grades of insulation and the four common magnetic materials.

Grade	Max Temp. (°C)	SmCo	NdFeB	Alnico	Ferrite
A	105	1.41	1.54	1.33	1.84
B	130	1.55	1.73	1.43	2.24
F	155	1.70	1.98	1.56	2.74
H	180	1.85	2.27	1.67	3.36

Thus, depending on the maximum operating temperature and the magnetic material, the mechanical time constant can increase as much as 40 to 336% above the "normal" 25 °C value. This results in the mechanical break point moving down in frequency with a resultant increase in the low frequency phase shift and decrease in relative stability.

Of the two effects, the mechanical is the more important and the expected temperature rise of the system being designed should be factored into the analysis of system stability.

Consideration of these temperature effects could greatly impact the selection of the motor size and its magnetic material

3.1.6.9 Motor/Gearhead Assembly

With a motor alone on a thermally passive surface, the winding temperature is determined only by the motor dissipation and the thermal characteristics of the motor/mount combination.

With a motor/gearhead assembly, the gearhead is typically located between the motor and the mounting surface. The motor dissipation must "flow through" the gearhead to reach the mounting surface. Two effects can take place:

- Since the gearhead adds additional surface area to the combination, the effective winding to ambient thermal resistance could decrease.
- The gearhead itself has dissipation and, therefore, the total combined dissipation could actually result in an increase in winding temperature depending on the operating velocity and torque level.

It is difficult to determine the final temperature and the effect on the rating of the motor in a motor/gearhead assembly as a result of the combination of these two effects using a purely theoretical analysis. Actual hardware testing is needed to determine which effect will predominate in a specific assembly.

The following data are the results of such testing of three sizes of brushless motor/gearhead assemblies in which each motor was mated with a gearhead of the same case size and power rating as the motor. The assemblies were mounted on the usual aluminum test plate and operated at maximum rated power levels, while monitoring torque, until acceptable steady state elevated temperature (155 °C winding temperature) was reached.

The data show a comparison between the continuous maximum torque rating of the motor alone and the motor when combined with a gearhead.

	Maximum Cont. Torque (N m)		
Motor Size (mm)	Motor Alone	Motor + Gearhead	% Change
60	0.544	0.621	+14
90	1.86	1.74	−6.5
115	4.65	3.66	−21

For the smallest size (60 mm), the first effect predominates and the result is an increase in available motor torque.

For the 90 and 115 mm sizes, the second effect predominates and the result is a decrease in available motor torque.

Although this test was of a limited nature, it would appear that, in general, larger motors will suffer degradation in torque capability when mated with a matching gearhead due to the gearhead adding to the total dissipation without a corresponding increase in surface area, resulting in an increase in the effective thermal resistance of the motor.

3.1.7   Motor Constant

Various factors are used to describe a motors capability and to compare motors when choosing the best one for an application.

Among these factors are efficiency, power rate (see Section 4.5.2) and motor constant.

Efficiency is difficult to calculate due to the lack of information about various losses, such as velocity-dependent losses, friction losses, and so on, in addition to I^2R losses.

Power rate is somewhat predictable, but is typically given in catalog data with respect to operation of the motor alone (no load inertia or torque load) using the one time peak torque rating of the motor in order to somewhat exaggerate the motors' performance.

The motor constant was originally created to describe the efficiency of special brush motors, known as "torquers" that create torque at zero velocity. Since there is no motion there is no output power; therefore all torquers have zero efficiency and zero power rate, as usually defined, eliminating those factors for describing performance and for comparison purposes.

However, a torquer does have input power (I^2R) and a performance factor was created, namely the ratio of torque to I^2R, to describe torquer efficiency;

$$K_{M(torquer)} = \frac{T}{I^2R} \tag{3.57}$$

For motors, both brush and brushless, the motor constant has been modified to be the ratio of torque to the square root of I^2R as follows:

$$K_M = \frac{T}{\sqrt{I^2R}} = \frac{K_T I}{\sqrt{I^2R}} = \frac{K_T}{\sqrt{R}} \; \text{N m W}^{-1/2} \tag{3.58}$$

This is a universal description of motor performance since K_M becomes independent of how a particular motor is wound, as long as the copper volume remains unchanged as the number of turns in the winding is changed. For example, if the number of turns is increased then the wire cross-section must be decreased by the same proportion, keeping the copper volume and the motor constant unchanged.

The motor constant for various DC motors depends on their construction and type of excitation.

3.1.7.1   Brush Motors

$$K_M = \frac{K_T}{\sqrt{R}} \tag{3.59}$$

3.1.7.2 Brushless Motors

Since brushless motors can be configured as either Wye or Delta, it is best to express the motor constant on a per phase basis and in accordance with the method of excitation:

- Sine BEMF with Sine excitation:

$$K_{M-S,S} = \left(\sqrt{\frac{3}{2}}\right)\left(\frac{K_{EP}}{\sqrt{R_P}}\right) \tag{3.60}$$

- Trapezoidal BEMF with Six Step excitation

$$K_{M-T,6} = \left(\sqrt{3}\right)\left(\frac{K_{EP}}{\sqrt{R_P}}\right) \tag{3.61}$$

- Sine BEMF with Six Step excitation.

$$K_{M-S,6} = \left(\frac{9}{2\pi}\right)\left(\frac{K_{EP}}{\sqrt{R_P}}\right) \tag{3.62}$$

$$K_{EP} = \text{BEMF constant peak in V rad}^{-1}\,\text{s}^{-1}$$

$$R_P = \text{phase resistance}$$

However, notice that since the expressions for K_M do not contain a velocity term they essentially measure the stall torque capability of the motor. Therefore, although useful for rating and comparing motors for their stall torque capability, additional data must be examined to make a proper choice as shown in the following.

A Wye connected brushless motor with sine BEMF and sine drive has the following specifications;

$$K_{E,L-L} = 0.39 \text{ V rad}^{-1}\text{s}^{-1} \qquad K_{T,L-L} = 0.39 \text{ N m A}^{-1} \qquad R_{L-L} = 2.6\ \Omega$$
$$\theta = 1.06\ ^\circ\text{C W}^{-1} \qquad P_C = 2\text{W @ 1000 rpm} \qquad \text{Wdg temp} = 155\ ^\circ\text{C max}$$

- At stall, with winding at 155 °C in a 25 °C ambient, the maximum allowable dissipation is:

$$I^2 R = \frac{155 - 25}{1.06} = 122.6 \text{ W} \tag{3.63}$$

From $P = \left(\frac{3}{2}\right) I_P^2 R_P$ for the dissipation in a three phase sine BEMF/sine drive,

$$I_P = \sqrt{\frac{2 \times 122.6}{3 \times \dfrac{2.6 \times 1.5}{2}}} = 6.48 \text{ A} \tag{3.64}$$

From $T = \left(\frac{3}{2}\right) K_p I_p$ for the developed torque in a three phase sine BEMF/sine drive,

$$T = \left(\frac{3}{2}\right) \left(\frac{0.39}{\sqrt{3}}\right) (6.48) = 2.19\,\text{N m} \tag{3.65}$$

$$K_{M-S,S} = \frac{2.19}{\sqrt{122.6}} = 0.197\,\text{N m W}^{-1/2} \tag{3.66}$$

- At 6000 rpm, the $I^2 R$ must be reduced by the velocity-dependent core loss in order to maintain the winding temperature at 155 °C.

$$I^2 R = \frac{155 - 25}{1.06} - P_C = 122.6 - (2)(6)^{1.5} = 93.3\,\text{W} \tag{3.67}$$

$$I_P = \sqrt{\frac{2 \times 93.3}{3 \times \dfrac{2.6 \times 1.5}{2}}} = 5.65\,\text{A} \tag{3.68}$$

$$T = \left(\frac{3}{2}\right) \left(\frac{0.39}{\sqrt{3}}\right) (5.65) = 1.91\,\text{N m} \tag{3.69}$$

$$K_{M-S,S} = \frac{1.907}{\sqrt{93.3}} = 0.197\,\text{N m W}^{-1/2} \tag{3.70}$$

Note that the torque capability dropped 13% due to operation changing from zero velocity to 6000 rpm while the motor constant did not change.

In a similar fashion, if the thermal resistance changed from 1 to 2 °C W^{-1}, the torque would drop to 1.6 N m at stall and 1.18 N m at 6000 rpm while the motor constant would remain at 0.197 N m W$^{-1/2}$.

3.1.8   Linear Motor [7–10]

Linear motors can eliminate the critical velocity, backlash, belt stretch, compliance and frictional wear problems associated with lead screw, belt and pulley and rack and pinion systems. They are available in a number of configurations as follows.

3.1.8.1   Flat Linear Motor – Iron

The development of the rare earth magnet based brushless motor has led to the design of the flat linear motor. If the housing of a brushless motor is split on one side and laid flat and the rotor is also flattened, the result will be as shown in Figure 3.29b.

This is an exact linear version of the rotary brushless motor in which the magnet assembly, also called the forcer, slider or glider, is the moving member. In this version, since the coil assembly is mounted on the machine structure, thermal conductivity is low.

The forcer has no dissipation or cabling allowing for an efficient attachment to the machine moving component.

MOVING COIL ASSEMBLY

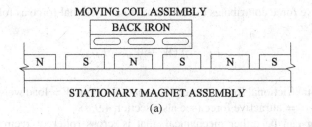

STATIONARY MAGNET ASSEMBLY
(a)

MOVING MAGNET ASSEMBLY

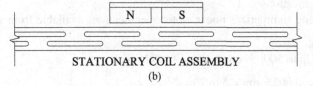

STATIONARY COIL ASSEMBLY
(b)

Figure 3.29 Linear motor schematics

Unfortunately, this version does not lend itself to the creation of long stroke motors since extended coil assemblies would involve high labor and material costs and would not lend itself to modular design.

Instead, a flat version of the normal brush motor is used, in which a three phase coil assembly is the moving member as shown in Figure 3.29a.

This arrangement results in the ability to make the stationary member as a modular section such that any number of them can be placed end-to-end to create extremely long travel, virtually unlimited in length.

A negative attribute of this configuration is that the forcer coils and Hall sensors must be connected to the controller via a cable that moves with the forcer. If not properly implemented, the cable can create a non-linear drag force on the forcer. In addition, for designs in which air or water cooling is used, there air hoses that will be part of the cable:

- Dynamics
  A linear system dynamics can be described by

$$F = mS'' + BS' + F_L + F_F \qquad (3.71)$$

where F = total force, B = viscous damping, F_L = load force, F_F = frictional force, m = system mass, S'' = linear acceleration, and S' = linear velocity.

The acceleration term is usually written as $W(\frac{S''}{g})$ where: W = system weight, g = acceleration due to gravity, leading to the term, "number of gs" of acceleration.

For example, if a system requires 2gs, ($S'' = 2g$) it needs a force of 2 times its weight.

- Attractive Force
  Due to the back iron on the forcer, there is a high attractive force between the forcer and the stationary member, which can be as high as 15 times the maximum continuous force rating of the motor. This in turn places a high load on the bearing structure of the motor/load assembly which actually contributes to the stiffness of the bearings, resulting in a stiffness of up to ten times that of an equivalent ball screw system.

The attractive force contributes to the normal total frictional force as follows:

$$F_F = \mu \left(W + W_M + F_{MA} \right) \tag{3.72}$$

where F_F = total frictional force, μ = coefficient of friction, W = load weight, W_M = motor weight, and F_{MA} = attractive force (see also Section 4.9.8).

The bearing can be either mechanical, that is, cross roller or recirculating, or non-mechanical, that is, air bearing or magnetic bearing.

- Nominal Specifications

  The following summarizes linear motor specifications available from various manufacturers.

    Force: Up to 5000 N continuous; 1400 N peak

    Velocity: 0.0025 mm s^{-1} to 7 m s^{-1}

    Acceleration: Up to 20 gs

    Stroke: 0.25 mm to 8 m (virtually unlimited)

    Accuracy: Down to sub-micron with appropriate encoder or laser interferometer.

- Feedback

  Hall sensors, or their equivalent, are required for the commutation of the three phase forcer coils.

  For precise motion and sine wave generation a linear encoder, with the glass scale length equal to the maximum travel is required.

  A laser interferometer can be used for ultimate resolution.

- Velocity/Controller BW/Resolution

  The velocity is limited by the controller bandwidth and the feedback sensor resolution according to the following:

$$\text{Velocity} = \text{Bandwidth} \times \text{Resolution} \tag{3.73}$$

For example, if the controller can process signals no faster than 5 MHz and the encoder has a resolution of 0.2 μm the maximum usable velocity is 1 m s^{-1}.

A dual mode design will provide for increased slew velocity by using the Hall sensors in the slew mode and the encoder when homing in to zero velocity

- Thermal [11]

  The coils on the forcer will have a maximum specified absolute temperature limit of 100 to 130 °C. There must be a temperature sensor mounted in the forcer coils to allow monitoring by the controller. Heat can be removed from the coils in a number of ways:

    conduction: heat can flow from the forcer into the equipment on which the motor is mounted

    radiation: a finned heat sink or plate of specified size can be mounted on the forcer

convection: air at the required pressure or water at the required inlet temperature and flow rate can be supplied to the forcer.

The conduction and radiation are considered to be "no cooling" while convection cooling is classified as "AC" or "WC". Air and water cooling can produce an increase in the continuous force rating:

Air: 12 to 25%

Water 25 to 85%.

- Cogging

  There will be cogging similar to that in a rotary motor which is not specified in product data sheets but should be discussed with potential suppliers to determine its effect on a particular application. Motor manufacturers use various methods to minimize cogging such as:

  skewing the magnets

  shaping the magnets

  skewing the forcer lamination slots.

- Air Gap and Flatness.

  The air gap between the forcer and the magnet assembly is critical in establishing the proper level of the magnet field strength. The bearing structure and the flatness of the surface on which the magnet assembly is mounted must be tightly controlled. For example, a variation of 30% in the air gap can produce a 5% variation in the force constant that can create a low frequency force constant ripple determined by the pitch of the flatness variation.

3.1.8.2 Flat Linear Motor – Ironless

By mounting two sets of linear magnet assemblies opposite each other, as shown in Figure 3.30, a "U"-shaped structure is created. The forcer is located in the air gap between the two magnet assemblies.

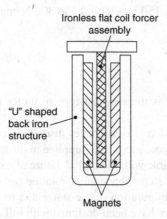

Ironless flat coil forcer assembly

"U" shaped back iron structure

Magnets

Figure 3.30   Ironless linear motor schematic

The forcer, consisting of a molded three phase coil set, contains no iron, resulting in zero cogging. The forcer design is analogous to the printed circuit disc and basket weave armatures in low inertia rotary motors. In addition, there is no magnetic attraction force as in the iron motor, reducing bearing loading.

The result is a forcer with lower mass than a comparable iron based forcer, providing high acceleration capability. The lack of iron however results in the ironless forcer having a higher thermal resistance than an equivalent iron forcer.

3.1.8.3   Tubular Linear Motor

Tubular linear motors have a cylindrical rod, consisting of a stack of axially magnetized disc magnets alternating with soft iron discs, located in the center of a hollow cylinder containing a set of three phase coils. The rod is supported by bearings at each end of the cylinder. When the coils are energized per normal three phase drive power, the rod will move through the cylinder, exerting a force proportional to the motor's force constant and the applied current. The action is similar to a solenoid.

Hall sensors and a linear encoder provide for coil commutation and closed loop positional control as in the flat brushless iron and ironless motors.

Typical stroke length is 100 to 500 mm with up to 4000 N peak force.

The design can result in a detent force of 5 to 15% of the peak force rating which can be used as form of holding brake.

3.1.8.4   Voice Coil Linear Motor

The voice coil linear motor, or actuator, is an adaptation of the device used to actuate loud-speaker cones, hence the name "voice coil".

It consists of coil of wire mounted within a radially magnetized magnet field. Control is achieved by simply applying current to the coil in the proper polarity to move the coil in or out of the field. The assembly is cog free, has no hysteresis and with a linear encoder can provide high acceleration/high frequency response closed loop performance.

Typical stroke length is 3 to 150 mm with 0.3 to 300 N continuous force and 0.8 to 700 N peak force.

3.1.9   Stepper Motors [12]

In concept, the stepper motor is the simplest type of motor from both its construction and control viewpoint.

It basically operates open loop [13] and takes discrete steps as the current supplied to its stator coils is controlled in response to pulses supplied to its drive electronics from the system controller. It is ideally compatible with the digital nature of modern control systems.

The basic magnitude of the steps (the angular rotation per step), the so-called full step, is created by the mechanical configuration of the stator and rotor; the number of poles in the stator and rotor. Stepper motors have been designed with full steps of a wide range of values such as 30, 15, 7.5, 1.8, 0.9, 0.45°, and so on.

The velocity and direction of the stepper motor is determined by the frequency and timed sequence of the command pulses, respectively, while the total distance traveled is determined by the number of command pulses. It can be controlled to follow the acceleration ramps and trapezoidal profiles typically used in modern motion control systems. The total distance traveled is equal to the total number of pulses commanded scaled by the angular resolution of the motor.

At zero velocity the stepper motor, if operated within its dissipation limit, can develop a continuous holding torque independent of load dynamics up to its maximum rating.

There are however two basic negative aspects of stepper motor technology:

- Due to inductive effects, there is a rapid fall off of torque at speeds above 2000 rpm
- The motor itself is a highly under damped electro-mechanical assembly exhibiting relatively low frequency resonance.

Both of these items have been addressed in literally hundreds of articles concerning how to design the control electronics to minimize these effects and extend the capability of stepper motors, including the addition of encoder feedback, which to a certain extent negates the idea of stepper motors providing low cost open loop control.

A detailed review of these various techniques is beyond the scope of this book but some are mentioned in the following sections in which the basic operation of the three most popular stepper motor designs is reviewed.

3.1.9.1   Variable Reluctance Stepper Motors

The variable reluctance stepper motor is the oldest design and operates on the principle of an electromagnet attracting a ferrous material (soft iron) to minimize the air gap between them, that is, reducing the reluctance to a minimum.

Figure 3.31 shows a cross-section of a typical VR stepper motor. The stator has four wound pole pairs (A–A, B–B, C–C and D–D) and is, therefore, designated as a four phase motor. The rotor has six salient poles. Motors are also available as three and five phase designs.

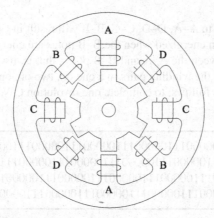

Figure 3.31   Stepper motor cross-section; variable reluctance

As shown, with poles A–A energized, two poles of the rotor have been attracted to the A–A poles. If the A–A poles are next de-energized and the B–B poles are energized then the rotor will rotate CW until the next pair of rotor poles, located CCW from the first pair, line up with the B–B poles.

The pitch of the stator poles is 360/8 = 45°

The pitch of the rotor poles is 360/6 = 60°

Therefore, the rotation per step will be 60–45 = 15°

The total number of steps per revolution = 360/15 = 24 steps per revolution.

In general, if:

$$P_S = \text{the number of stator poles}$$

$$P_R = \text{the number of rotor poles}$$

$$N = \text{number of steps per revolution}$$

then

$$N = \frac{P_S \times P_R}{P_S - P_R} \tag{3.74}$$

By energizing the four winding pairs sequentially per the following schedule, the motor will rotate in the full step mode in which 24 steps will result in one revolution CW

Wdg A–A	100010001000100010001
Wdg B–B	010001000100010001000
Wdg C–C	001000100010001000100
Wdg D–D	000100010001000100010

Changing the sequence to A–A, D–D, C–C, B–B will result in CCW rotation.

If the A–A poles remain energized when the B–B poles are energized, the rotor will rotate to a position midway between the A–A and B–B poles, that is, it will rotate 7.5° or one half a step CW By sequencing the winding pairs in a one on, two on, one on, and so on pattern, it will take 48, 7.5° steps, as follows, to complete one revolution CW.

Wdg A–A	110000011100000111000001110000011100000111000011
Wdg B–B	011100000111000001110000011100000111000001110000
Wdg C–C	000111000001110000011100000111000001110000011100
Wdg D–D	000000111000001110000011100000111000001110000110

An additional full step sequence is created by energizing two windings in rotation as follows:

Wdg A–A	1001100110011001100110011
Wdg B–B	1100110011001100110011001
Wdg C–C	0110011001100110011001100
Wdg D–D	0011001100110011001100110

In this sequence the full steps will be shifted 7.5 °C W from the positions achieved in the sequence with only one coil on per step.

By energizing two windings at a time, higher torque will be obtained at the expense of doubling the motor current.

Note that in the prototypical VR motor, neither the stator nor rotor poles have a specific polarity. The rotor will be attracted to the stator poles whether, for example, the A–A poles are N–S or S–N.

3.1.9.2 Permanent Magnet Stepper Motors

The permanent magnet stepper motor is similar to the VR motor with the salient pole rotor replaced by a cylindrical assembly of permanent magnets, as shown in Figure 3.32.

In the PM motor, the stator pole polarities have to be designed to be compatible with the polarity of the rotor magnets.

For example, in Figure 3.32, pole A must be a south pole and A' must be a north pole to achieve the position shown.

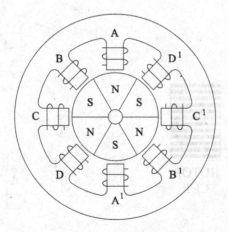

Figure 3.32 Stepper motor cross-section; permanent magnet

To rotate 15 °C W, poles A and A′ must be turned off and pole B must become a north pole, pole B′ a south pole.

The pole polarities and step sequence for a full step 360 °C W rotation is:

A	S000N000S000N000S000N000S
A′	N000S000N000S000N000S000N
B	0N000S000N000S000N000S000
B′	0S000N000S000N000S000N000
C	00S000N000S000N000S000N00
C′	00N000S000N000S000N000S00
D	000N000S000N000S000N000S0
D′	000S000N000S000N000S000N0

Half step and two phases on full step can be achieved in the same manner as for the VR motor.

3.1.9.3   Hybrid Stepping Motor

The hybrid stepping motor is essentially a combination of the VR and PM motors.

Figure 3.33 shows both a motor cross-section and side view of the rotor.

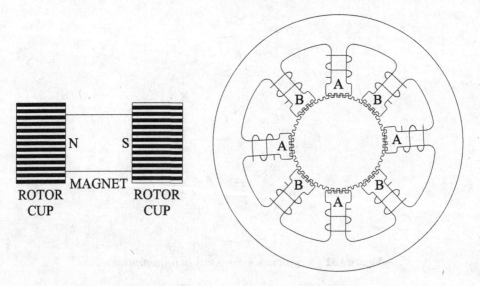

Figure 3.33   Stepper motor cross-section; hybrid

The cylindrical rotor consists of three main parts:

A center section that is an axially magnetized permanent magnet plus two end caps that have small salient poles, or teeth, around their circumference. This construction results in one end cap becoming a south pole and the other a north pole. The teeth on one end cap are offset from the teeth on the opposite end cap by one half a tooth pitch.

The stator has wound salient poles with teeth on the same pitch as the rotor teeth. The stator poles are divided into two phases. These two phases (A and B) of stator poles are mechanically offset by one-quarter of a tooth pitch.

The result is that when one phase (A) is energized, one of the rotor end caps will be in alignment with the north poles of the stator and the opposite end cap will be in alignment with the south poles of the stator. At this position the end caps will be within one-quarter of a tooth pitch of being in alignment with the unenergized phase (B).

When the phase excitations are changed (A off, B on) the rotor will then rotate one-quarter of a tooth pitch and be in alignment with the north and south poles of the B phase.

This then places the rotor within one-quarter of a tooth pitch of the A phase and when excitation is again changed (B off, A on) with the A polarity opposite of the original condition, an additional one-quarter tooth pitch of rotation will occur.

Continued alternate excitation of A and B with alternating polarity will then result in continued rotation of one-quarter of a tooth pitch per step.

Example

If the rotor has 50 teeth, the tooth pitch will be 7.2°. Therefore each step will result in 1.8° of rotation

This is the operation of the widely used 200 steps per revolution, two phase hybrid stepper motor.

3.1.9.4   Coil and Drive Configurations

The stator coils in stepper motors exist in a number of different configurations, depending on the type of motor and the use of either a bipolar or unipolar drive circuit.

Figures 3.34 and 3.35 show the main type of windings and circuits as follows:

- 4 wire: Two phases, two coils; two leads per coil bipolar drive Figure 3.35a
    Used with permanent magnet and hybrid motors. Requires "H" drive amplifier capable of sequentially reversing coil current to reverse stator field polarity.
- 6 wire: Two phases; two coils with each coil center tapped; six leads unipolar drive Figure 3.35b.

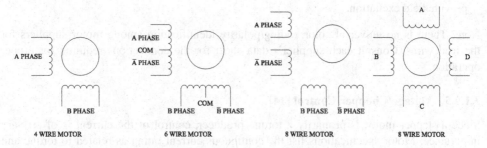

Figure 3.34   Stepper motor coil configurations

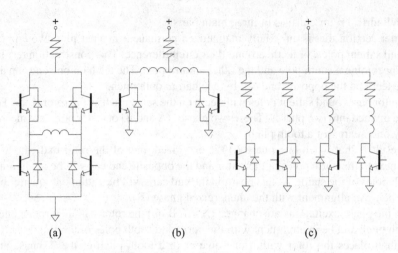

(a)   (b)   (c)

Figure 3.35   Stepper motor drive configurations

Used with permanent magnet and hybrid motors. The center taps are connected to the supply voltage and the coil end wires are sequentially connected to ground by the amplifier to reverse stator field polarity

(In a variation, the center taps are connected internally and brought out as a single lead – 5 wire)
- 8 wire: Two phases; two coils per phase; eight leads.

   Unipolar drive/bipolar drive (Figures 3.35(a) (b))

   Used with permanent magnet and hybrid motors.

   With the coils in each phase in series and the common connected to the supply voltage, a 6 wire configuration is created.

   With the coils in each phase in series and the common not terminated, a 4 wire high voltage/low current configuration is created.

   With the coils in each phase connected in parallel, a 4 wire low voltage/high current configuration is created.
- 8 wire: Four phases; one coil per phase; eight leads.

   Unipolar drive (Figure 3.35c)

   Typical four phase variable reluctance configuration

   One end of each coil is connected to the supply voltage

   The opposite end of each coil is sequentially connected to ground by the amplifier to provide field excitation.

Note: There is no universal color coding/polarity identification among motor suppliers for the lead wires. Consult each supplier's data sheet for the connections required for proper operation.

3.1.9.5   Voltage/Chopper Control [14]

Since a stepper motor is primarily a torque producer, control of the current is of primary importance. Motor specifications list the continuous current rating as related to torque and limited by maximum dissipation.

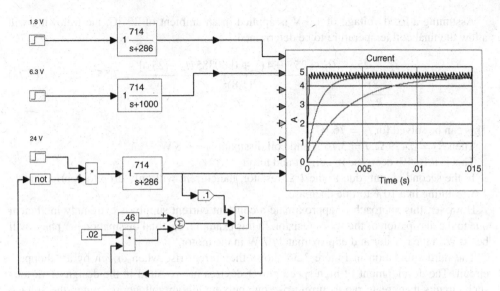

Figure 3.36 Coil current rise for three drive conditions

Rated voltage is then simply listed as the product of current and resistance, which typically results in a fairly low value.

Example
A motor with the following rating:

$$I = 4.5 \text{ A} \quad R = 0.4\,\Omega \quad L = 1.4 \text{ mH} \quad E = 1.8 \text{ V} \quad t_{\text{wdg}} = 130\,°C \text{ max} \quad t_{\text{amb}} = 50\,°C \text{ max}$$

Figure 3.36 shows the current rise in the winding for three drive circuits.

For the first, the winding is supplied by a constant 1.8 V (the rated voltage)

Note that for this condition, the TC_E equals 3.5 ms and it takes 14 ms for the current to rise to its final value. This will place an extremely low limit on the full torque stepping rate.

The second simulation in Figure 3.36 shows the current rise with a 1 Ω resistor placed in series with the winding and supplied by a constant 6.3 V.

Initial control circuits used this method to decrease the electrical time constant.

The TC_E has been reduced to 1 ms and the current reaches final value in 4 ms, increasing the full torque stepping rate by a factor of 3.5.

Although these two circuits are fairly simple they have two negative effects; the first is their low rise time, the second is the fact that since they are powered from a constant voltage, the current and, therefore, the torque will decrease as the resistance increases due to its dissipation.

For example, in the first circuit the effective thermal resistance can be derived from the rating data and the use of Equation 6.22 as follows:

$$R_2 = 0.4\,[1 + 0.00385\,(130 - 50)] = 0.523\,\Omega \tag{3.75}$$

$$I^2 R = 4.5^2 \times 0.523 = 10.6\,\text{W} \tag{3.76}$$

$$\theta = \frac{130 - 50}{10.6} = 7.55\,°C\,\text{W}^{-1} \tag{3.77}$$

Assuming a fixed voltage of 1.8 V is applied in an ambient of 25 °C, the following will allow the final coil temperature to be determined:

$$\frac{t_2 - 25}{\frac{(1.8)^2}{R_2}} = \frac{(t_2 - 25)\,[0.4\,(1 + 0.00385\,(t_2 - 25))]}{(1.8)^2} = 7.55 \tag{3.78}$$

This can be solved for $t_2 = 76\,°C$

Then $R_2 = 0.479\,\Omega$, $I = 3.76$ A and coil dissipation = 6.8 W.

This is a 16.4% decrease in current and torque.

In the second circuit, due to the 1 Ω resistor, the current will only drop to approximately 4 A resulting in a 10% torque decrease.

However, this approach to approximate a constant current supply is extremely inefficient due to the dissipation of the series resistor. In this example, the total dissipation per phase will be 16 W in the resistor and approximately 7 W in the motor.

The third simulation in Figure 3.36 shows the current rise when driven by a "chopper" circuit. The development of high speed power devices has resulted in the design of various such circuits that create rapid current rise by applying a high voltage, removing the voltage when the current exceeds the nominal value by a preset amount that then decays to a lower value as determined by the TC_E, and then reapplying the voltage.

In Figure 3.36 the current is set for 4.5 ± 0.10 A with a rise time of 0.3 ms.

The switching frequency is 3.8 kHz.

3.1.9.6   Series/Parallel Control

Eight wire motors can have the two coils in each phase connected either in series or parallel.

Example

A motor has coil ratings of: $R = 0.4\,\Omega$, $L = 2.7$ mH

in series: $I = 3.15$ A, $R = 0.8\,\Omega$, $L = 5.4$ mH, $E = 2.52$ V

in parallel: $I = 6.30$ A, $R = 0.2\,\Omega$, $L = 1.35$ mH, $E = 1.26$ V

Figure 3.37 shows the current for each connection when being driven by a chopper type circuit from a 48 V supply.

Note that in the parallel connection the current rises to the rated value in 0.19 ms compared to the series connection in which the current rises in 0.36 ms.

For the series connection	$I = 3.15 \pm 0.05$ A	with freq. = 4.3 KHz
For the parallel connection	$I = 6.30 \pm 0.10$ A	with freq. = 7.6 KHz

The result is that with the parallel connection, torque is maintained to higher speeds than with the series connection, as shown in Figure 3.38.

For torque = 1 N m	series speed = 600 rpm	parallel speed = 1200 rpm
= 0.6 N m	= 900 rpm	= 1800 rpm

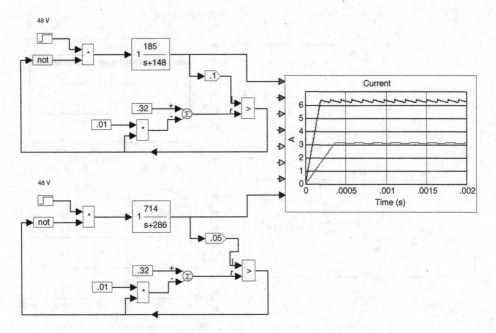

Figure 3.37   Chopper drive current for series and parallel coils

In addition the parallel connection provides "useful" operation to over 2000 rpm whereas the series connection is limited to 1200 rpm.

This example is shown for a particular motor but is illustrative of the fact that connecting the coils in parallel will typically give better performance than the series connection but will require twice the current from the driver circuit.

3.1.9.7   Variable Supply Voltage

The chopper control circuit simulated in Figures 3.36 and 3.37 shows how using a supply voltage much higher than the "rated" voltage can result in short current rise time leading to high stepping rates.

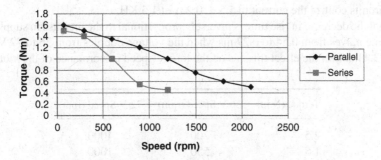

Figure 3.38   Speed/torque, series versus parallel connection

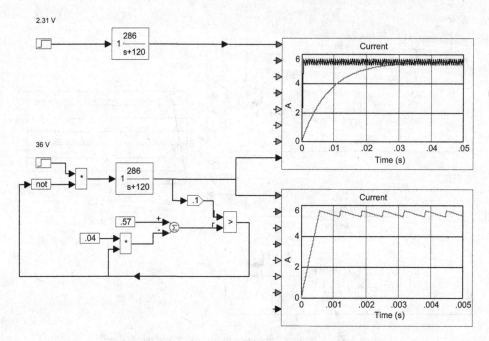

Figure 3.39   Chopper drive current for 36 V supply

It is interesting to investigate the effect of the value of the supply voltage on the rise time of the current and therefore on the torque versus speed of the motor.

Example

A motor with the following rating:

$I = 5.5\,\text{A} \quad R = 0.42\,\Omega \quad L = 3.5\,\text{mH} \quad E = 2.31\,\text{V}$
Inertia $= 0.143 \times 10^{-3}\,\text{kg m}^2$
$T_{\text{hold}} = 4.55\,\text{N m} \qquad T_{\text{detent}} = 0.127\,\text{N m} \qquad \theta = 2.65°\,\text{CW} \qquad t_{\text{amb}} = 40\,°\text{C max}$

Figure 3.39 shows the coil being driven by a chopper circuit from a 36 V supply and from a simple 2.31 V supply for comparison.

Figure 3.40 is the same except the chopper voltage has been changed to 72 V.

Both circuits control the current to 5.5 ± 0.2 A at 1.5 KHz.

Note that the decrease in rise time is inversely proportional to the increase in supply voltage; the rise time halves from 0.54 to 0.27 ms when the voltage doubles from 36 to 72 V.

Figure 3.41 shows the effect this has on the torque/speed performance of the motor.

Torque (N m)	36 V Speed (rpm)	72 V Speed (rpm)
2.5	200	4000
1.5	550	1000
1.0	900	1500

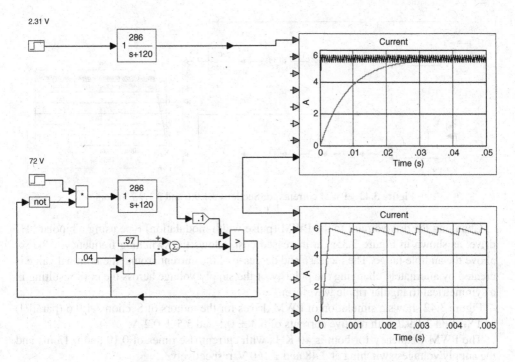

Figure 3.40   Chopper drive current for 72 V supply

3.1.9.8   PWM Control

The previous sections showed simulations of various chopper drive circuits, in which the switching frequency is determined by a combination of the TC_E and the upper and lower switching levels set for the current. To a certain extent this results in switching frequencies that are in the audible range. Some circuits use a timed current decay in order to increase the switching frequency and still maintain the lower cost of unipolar chopper topology with output stages as shown in Figure 3.35b.

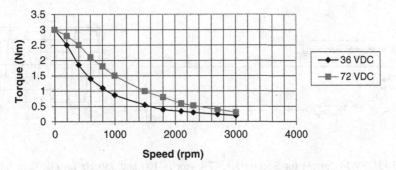

Figure 3.41   Speed/torque; 36 versus 72 V supply

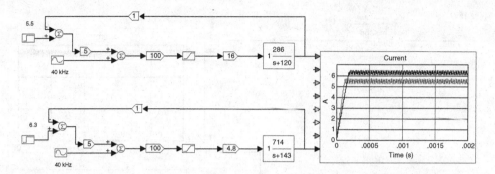

Figure 3.42 PWM currents for Sections 3.1.9.6 and 3.1.9.7 motors

Changing the drive circuit to the PWM (pulse width modulation) type using a bipolar "H" drive, as shown in Figure 3.35a, at increased cost, allows the switching frequency to be set above the audible range. The increase and decrease of the current around the nominal value is created by alternately changing the polarity of the supply voltage across the coil, resulting in a symmetrical triangular ripple waveform.

Figure 3.42 shows a simulation of PWM drives for the motors of Section 3.1.9.6 (parallel) and Section 3.1.9.7 with relative currents of 6.3 ± 0.1 and 5.5 ± 0.2 A.

The PWM frequency for both is 40 KHz with current rise times of 0.19 and 0.13 ms and the supply voltages switching at ±48 and ±160 V, respectively.

Figure 3.43 shows the coil current for the Section 3.1.9.7 motor when being driven at two different frequencies; that is, two different speeds. The motor is stepping at 200 full steps per revolution.

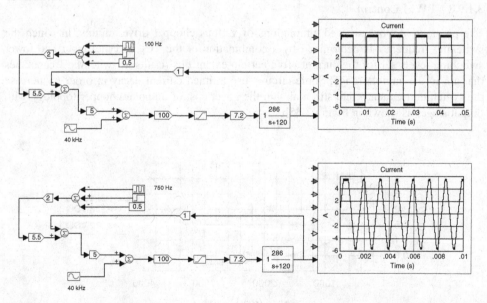

Figure 3.43 PWM current for Section 3.1.9.7 motor at 100 and 750 Hz (available in full color at www.wiley.com/go/moritz)

When being driven at 100 Hz, corresponding to 120 rpm, the current completes each cycle as a well defined square wave, indicative of the full torque capability of the motor, as shown by the 72 V speed/torque curve of Figure 3.41.

When being driven at 750 Hz, corresponding to 900 rpm, the current has deteriorated to a triangular wave that just peaks to the maximum value, the torque having fallen to approximately 50% of its maximum value, again as shown in Figure 3.41.

3.1.9.9 Resonance

One negative characteristic of all stepper motors is an under damped resonance each time a step command is given. This resonance is typically in the range of 100 to 300 Hz. It is similar to the response of an under damped spring–mass system In the stepper motor, the mass is the rotor and the "spring" is the magnetic field existing between the rotor and stator. Just as in the mechanical system, the frequency can be determined by:

$$f = \frac{1}{2\pi}\sqrt{\frac{K}{J}} \text{ Hz} \tag{3.79}$$

where K is the equivalent spring stiffness and J is the total inertia; rotor plus any connected load.

The motor of Section 3.1.9.6 (parallel) with:

$$T_{\text{hold}} = 1.78 \, \text{N m} \quad J_{\text{rotor}} = 5.7 \times 10^{-5} \, \text{kg m}^2$$

can be modeled and commanded to take half steps (0.9°) at a rate of 10 steps per second, as shown in Figure 3.44.

The motor exhibits the typical under damped response with a natural frequency of 240 Hz and ζ of 0.1. It takes approximately 50 ms to settle.

If the motor were to be given step commands at or about 240 steps per second it would become oscillatory and loose synchronism, that is, become unstable.

For this reason, unless some form of damping is implemented, stepper motor systems should not be driven in the range of their resonant frequency. To avoid resonance problems in variable velocity or positional systems, the motor should be commanded to "run through" the resonant frequency, implement one of the following damping methods or operated in the micro-stepping mode.

3.1.9.10 Damping

Various methods have been created to reduce or eliminate the resonance problem in stepper motors [15, 16]. New approaches to implement damping are continually being created by controller manufacturers and should be reviewed for suitability during the design process. Damping methods consist of two general categories, mechanical or electronic, some of which are:

(Note that in many of the electronic methods, information about shaft position in a finer resolution than the stepping increments is needed, which requires the use of an encoder, creating a form of closed loop control).

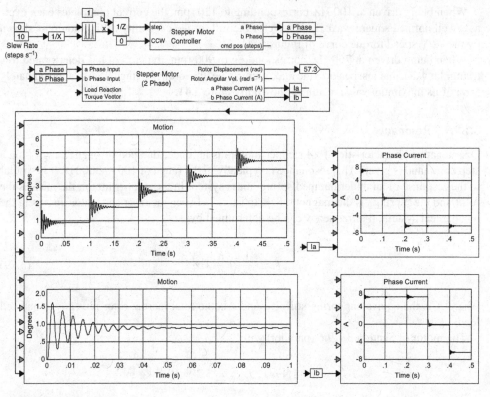

Figure 3.44   Section 3.1.9.6 motor at 10 steps/s, half stepping (available in full color at www.wiley .com/go/moritz)

- Mechanical
  - Addition of coulomb friction in the form of a drag brake. In Figure 3.45 coulomb friction of 0.0141 N m has been added, causing the settling time to decrease to approximately 25 ms.

    Adding friction has improved the operation but will also increase the torque load on the motor which will decrease the torque available to drive the load inertia.

    Also, the brake will be subject to wear, changing its' effect or requiring periodic adjustment.

    In Figure 3.46, the inertia has been increased to 11.4×10^{-5} kg m^2, that is, a condition with load inertia equal to the motor inertia.

    The settling time has increased back to 50 ms and the frequency has decreased to 168 Hz, in agreement with: $f^2 \propto \frac{1}{J}$, since $\left(\frac{240}{168}\right)^2 = 2$.

    To again reduce the settling time the friction would have to be increased further.
  - Addition of an inertial damper, typically to the rear of the motor, increasing the inertia and lowering the resonant frequency below the speed range of interest.
  - Addition of a viscous damper that produces a torque proportional to the motor velocity. Although very effective it will slow down the system response.

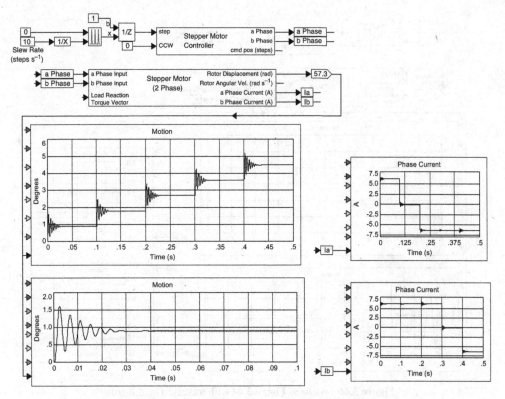

Figure 3.45   Same as Figure 3.44. with coulomb friction added

Figure 3.47 shows the response of the simulated system with the addition of $0.05 \, kg \, m \, rad^{-1} \, s^{-1}$ of viscous friction.

- Electronic [17–20].
  - A motor with two phases on for each step will typically have better damping than a single phase on, or the one on/two on type of drive.
  - A control in which switching from phase to phase is modified to operate in a fashion similar to a bang-bang time optimal servo system.

    Instead of a step being simply $A \rightarrow B$, the motion is "braked" for a short time before B is reached by reapplying power to A and then finally to B.

$$A \rightarrow B \rightarrow A \ (\text{for a short time}) \ \rightarrow B$$

  - A method of "delayed last step" in which the turn off of the next to last step is delayed until its overshoot brings the rotor close to the last step, at which time the next to last is turned off and the last is turned on.

    This method is limited to a minimum of two steps and actually depends on the existence of a certain amount of overshoot.

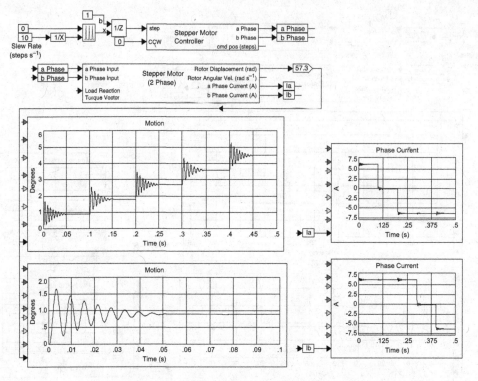

Figure 3.46   Same as Figure 3.44 with increased load inertia

- A method using two different voltages in which a high voltage is used during phase turn on and a low voltage is used during turn off. This method can only be used with the two phases on control where as one phase is turning off, the next phase is turning on.
- A method used in the one phase on control in which the drive circuit is designed so the current has a rapid turn on and a slow turn off, causing the turning off phase to have a "hold back" effect, which improves damping.

3.1.9.11   Microstepping [21]

As noted in Equations 3.1 and 3.2, the basic relation between torque and current in a magnetic field is:

$$T = BIlr \sin\theta \tag{3.80}$$

This can be written as:

$$T = IK_t \tag{3.81}$$

where

$$K_t = Blr \sin\theta = K_x \sin\theta \tag{3.82}$$

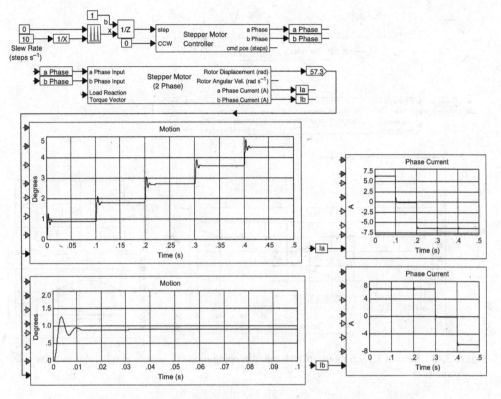

Figure 3.47   Same as Figure 3.44 with addition of viscous damping

which is a function of the physical design with a sinusoidal variation ($\sin \theta$) in each magnetic cycle.

If the applied current is a sine wave, then:

$$T = (I \sin \theta)(K_x \sin \theta) = IK_x \sin^2 \theta \tag{3.83}$$

If this represents the torque developed on one phase (phase a) of a two phase structure, then:

$$T_a = IK_x \sin^2 \theta \tag{3.84}$$

Similarly, if phase b is 90° out of phase with a, it will have a cosine relation such that:

$$T_b = I K_x \cos^2 \theta \tag{3.85}$$

The total torque: $T_t = T_a + T_b = IK_x (\sin^2 \theta + \cos^2 \theta) IK_x$ (since $\sin^2 \theta + \cos^2 \theta = 1$)

$$\tag{3.86}$$

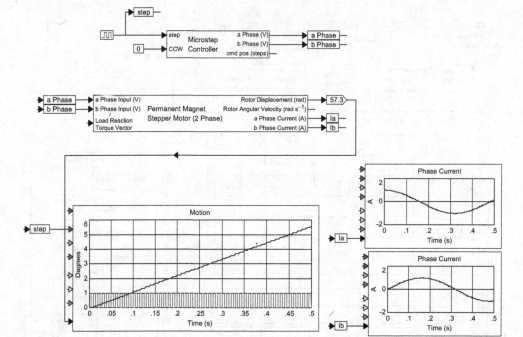

Figure 3.48 Section 3.1.9.6 motor in 1/16 microstepping mode (available in full color at www.wiley .com/go/moritz)

The result is a two phase synchronous motor with constant ripple-free torque.

If the amplitudes of the sine and cosine currents applied to the motor are created by "stepping along" in discrete fractions of the maximum amplitude, the rotor will follow along in discrete steps, or microsteps.

Controllers are available in microsteps as low as 1/256, which in a 200 full steps per revolution creates a resolution of 0.007 031 25° per microstep.

Figure 3.48 shows the motor taking 1/16 microsteps (0.1125° per step.) at a rate of 100 steps per second, which in 4 s results in a move of 4.5°.

This is the same move distance and time shown in Figures 3.46 and 3.47 when taking half steps (0.9° per step) with square wave drive.

Note the smooth motion and lack of resonance caused by the incremental step by step torque created by the stepped sine and cosine currents.

A negative aspect of icrostepping is that the incremental torque per microstep is a fraction of the full step holding torque. It can be shown that [22]:

$$T_{\mu s} = T_{fs} \left(\sin \frac{90}{\frac{\mu s}{fs}} \right) \qquad (3.87)$$

where:

$T_{\mu s}$ = torque per microstep T_{fs} = full step torque $\dfrac{\mu s}{fs}$ = microsteps per full step

In the simulated example; $\dfrac{\mu s}{fs} = 16$ $T_{\mu s} = 1.78 \sin 5.625 =$
$T_{fs} = 1.78$ N m 0.174 N m

In this case, if the sum of all load torques is greater than 0.174 N m, a single step command will result in no motion! As the number of microsteps per full step increases, this effect will multiply. It may take a number of microstep commands before the torque increases to a value large enough to initiate motion. Although microstepping increases theoretical resolution, this effect must be considered when determining accuracy.

An alternate method to achieve a form of microstepping is to marry the stepper motor to a reduction gearhead. For example if the motor in the previous simulation were connected to a 16:1 reduction gearhead, the output shaft would rotate in 0.1125° increments with 16 times the motor full step torque when the motor rotates in 1.8° full step increments.

This method does have a drawback in that the output velocity is reduced by the gearhead ratio, limiting this approach to "low speed" applications.

3.1.10   Induction Motors

On May 1, 1888 US patent No. 381 968 was issued to Nicola Tesla (1857–1943) a brilliant but somewhat eccentric engineer and inventor.

The patent describes a unique motor design capable of:

> "... effecting a progressive shifting of the magnetism"

which is today better known as a "rotating magnetic field" and essentially revolutionized the field of powered motion, rapidly obsolescing steam power and advancing the industrial revolution. As an aside: Tesla sold the patent to George Westinghouse and together they established AC as the basic method of generating, transmitting and using electrical energy in the United States and eventually over the world.

The rotating magnetic field is the basic effect in the operation of the induction motor.

3.1.10.1   Motor Construction

Figure 3.49 shows a cross-section of an induction motor consisting of a laminated wound stator and a "squirrel cage" rotor.

The stator has three groups of windings (A, B and C) powered by a three phase supply.

The three winding groups consist of sets of poles interleaved and equally spaced around the stator.

For example, a two pole motor will actually have six salient poles, two for each phase, as shown in Figure 3.50. A four pole motor will have twelve salient poles, and so on.

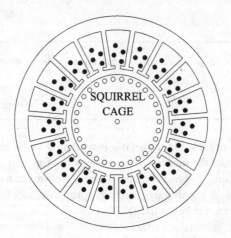

Figure 3.49   Squirrel cage induction motor cross-section

The rotor, shown in Figure 3.51, consists of a laminated steel cylinder into which are embedded a series of copper or aluminum conductors that in turn are terminated by conductive rings at each end of the assembly. The end rings essentially short circuit the conductors.

The conductors are skewed in order to:

- Decrease harmonic and ripple effects
- Eliminate locking if the number of stator teeth and rotor teeth are equal.

3.1.10.2   Basic Operation

When the stator windings are powered by the three phase supply, as shown in Figure 3.50, the poles will become North or South polarities at the respective times within a complete three phase cycle and create flux directions as follows:

Time	N	S	Total Flux Path
0	A_X/C_Y	A_Y/C_X	$1 \rightarrow 4$
60	B_X/C_Y	B_Y/C_X	$2 \rightarrow 5$
120	A_Y/B_X	A_X/B_Y	$3 \rightarrow 6$
180	A_Y/C_X	A_X/C_Y	$4 \rightarrow 1$
240	B_Y/C_X	B_X/C_Y	$5 \rightarrow 2$
300	A_X/B_Y	A_Y/B_X	$6 \rightarrow 3$
360	A_X/C_Y	A_Y/C_X	$1 \rightarrow 4$

The result is a flux pattern that rotates one mechanical cycle (one revolution) for each complete electrical cycle. Thus if the frequency is 60 cycles per second, the result will be 60

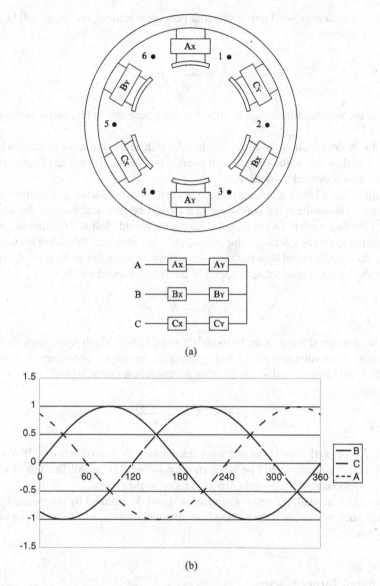

(a)

(b)

Figure 3.50 (a) Three phase pole and coil configuration, (b) three phase current waveshape timing

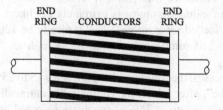

Figure 3.51 Squirrel cage rotor

revolutions per second or 3600 rpm. For a four pole arrangement, the speed will be 1800 rpm, and so on. In general:

$$N_S = \frac{120f}{p} \text{ rpm} \tag{3.88}$$

where N_S is the synchronous speed in rpm, f is the frequency in Hz, and p is the number of poles.

As the flux rotates around the rotor, a voltage is induced in the rotor conductors causing a high current to flow due to the short circuit created by the end rings. The effect is similar to a transformer with a shorted secondary.

The result is that a Lorentz force is created in each rotor conductor. In Equation 3.1, a force is created on a current-carrying conductor in a stationary flux field causing the conductor to translate. If the flux were to move, then the conductor would "follow" the flux or, in this case, if the flux rotates, the rotor will develop torque and start to rotate, following the flux.

Initially, the relative speed between the rotor and the rotating flux is N_S since the rotor speed is zero. As the rotor starts to rotate at speed N, the relative speed will be:

$$N_S - N \tag{3.89}$$

It is the relative speed between the rotor and the rotating flux which determines the magnitude and frequency of the voltage induced in the rotor and, therefore, the torque.

This effect can be described as the SLIP of the induction motor, where:

$$\text{SLIP} = S = \frac{N_S - N}{N_S} \tag{3.90}$$

At start, $N = 0$ and $S = 1$. As the rotor accelerates, N approaches N_S. If $N = N_S$ then $S = 0$ and the rotor speed would be equal to the rotating flux speed. But this is impossible, since there would be no voltage and current induced in the rotor.

Under no external load, the rotor will reach a speed determined by the no load conditions in the motor, such as bearing friction, windage, magnetic losses, and so on with a no load slip of approximately 2%.

3.1.10.3 Speed/Torque Curve

Unlike the speed/torque curve for a DC motor, which is linear from stall to no load, the speed/torque curve for an induction motor is extremely nonlinear as shown in Figure 3.52. The stall torque, called "locked rotor torque", will typically be 120 to 250% of the full load torque. The stall, or "locked rotor current", can be as high as 300 to 1000% of the full load current.

As the rotor speed increases, the torque will experience a 15 to 25% decrease to a low point, called the "pull up torque", after which the torque will continually increase until the "pull out torque", which typically occurs at 75% of synchronous speed, is reached. The torque then rapidly decreases to the no load value at a no load slip of approximately 2%.

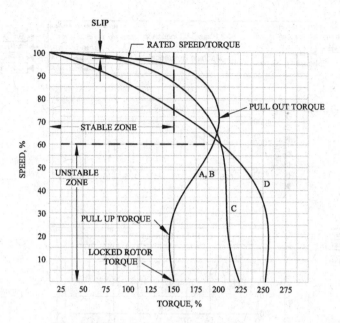

Figure 3.52 Induction motor speed/torque curves

During acceleration from stall to no load, the current will start at the "locked rotor current "value, remain at approximately that value until the speed has reached 75 to 80% of synchronous speed and then drop rapidly to the no load value. A typical start current transient is shown in Figure 3.53 for a simulated motor.

The speed/torque curve can be broadly divided into a stable zone and an unstable zone, as shown in Figure 3.52. In the stable zone, any increase in load torque will result in an increase in motor torque, counteracting the load torque and settling at a lower speed with a small increase in slip, and vice versa for a decrease in load torque.

In the unstable zone, any increase in load torque will result in a decrease in motor torque leading to continual decrease in speed until stall is reached.

The A, B, C and D curves are typical of the standard types of induction motors specified by the National Electrical Manufacturers Association (NEMA).

Detailed analysis of the polyphase motor shows that the developed torque is directly proportional to the rotor copper losses and inversely proportional to slip. By manipulating these parameters in the motor design, various speed/torque characteristics can be achieved.

3.1.10.4 Voltage/Frequency – V/Hz Control

Equation 3.88 shows a direct correlation between speed and applied frequency leading to early controllers being designed to achieve speed control by only varying the controller output frequency and voltage without any control over the voltage and current phase. Such controllers are also known as "inverter drives."

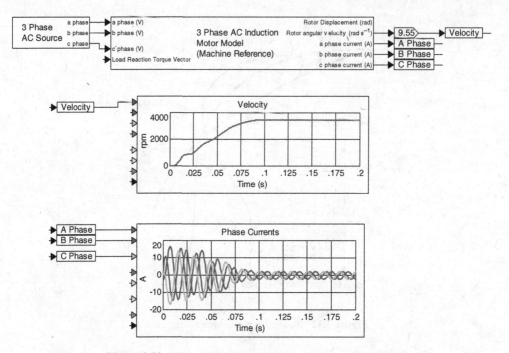

Figure 3.53   Three phase motor starting current and velocity

Lowering the frequency from the base value will lower the speed, but will also lower the effective reactance and the "counter emf" generated by the rotor rotation. If the applied voltage were to remain constant, the current would increase, resulting in an increase in the motor flux (see Section 3.1.10.5 for a description of the component parts of the input current).

Since the flux must remain constant, the voltage must be reduced as the frequency is reduced, resulting in the requirement that the ratio of voltage to frequency be kept at a nominal constant value.

This type of control can be used to provide "soft start" operation in which the starting current can be limited to no more than 150% of rated current by starting operation at low frequency and voltage, keeping slip low, and slowly increasing both until operating speed is reached. This essentially provides for a controlled acceleration/deceleration.

A problem with the "soft start" implementation is that insufficient torque will be generated to overcome load torque during low speed start. Some designs attempt to overcome this problem by providing for a "voltage boost", that is, more voltage than the V/Hz ratio would command.

This boost is implemented by trial and error and is set for a single value as a percentage of full load current. However, if set too high the result can be an over flux condition for light loads resulting in motor overheating, or if set too low insufficient flux for peak loading.

An additional problem with this type of control is with regard to transient response.

If the load torque is changed too rapidly, the circuitry will not correct the voltage and frequency fast enough to compensate. For example, if load torque increased in a step fashion, speed would drop (creating a step increase in slip) without a counteracting increase in developed torque and operation could move into the unstable part of the speed/torque region, resulting in a stall condition.

For these reasons, the V/Hz type of controller is most suited for constant or slowly varying loads providing simple, relatively low cost, speed control.

3.1.10.5　Vector or Field Oriented Control [23–26]

As noted in Section 3.1.2 and Figure 3.4, the two fields in a DC motor (B_F and B_A) are kept at a 90° spatial relation by the act of commutation, resulting in a maximum torque, independent of the armature velocity θ'. The torque can be controlled by the armature current separate from the velocity, which is controlled by the applied voltage.

For example, in a wound field DC motor in the steady state and ignoring resistive drops:

$$E_{\text{in}} = K_E \theta' = K_1 B_F \theta' = K_2 I_f \theta' \qquad \therefore \theta' = \frac{E_{\text{in}}}{K_2 I_f} \tag{3.91}$$

and $T = K_t I_a = K_3 B_F I_a = K_4 I_f I_a$ where K_n are constants dependent on motor design.

Thus, if I_f is constant, θ' will vary directly with E_{in} and T will vary directly with I_a, providing independent control of speed and torque.

In addition, maximum speed, the base speed θ'_{base}, will be reached when: $E_{\text{in}} = E_{\text{max}}$.

However, speed can be increased above the base speed by field weakening; reducing I_f.

The object of induction motor control is to emulate DC motor control by being able to perform these independent functions [27].

However, in the induction motor, the rotor voltage and the resulting current and flux, is transferred to the rotor by induction from the stator and, therefore, is not physically available for independent control. The rotor values must be controlled by manipulation of the magnitude and phase of the stator voltage and current.

This is accomplished by the use of the mathematical transformation of three phase AC currents into two imaginary DC values using transforms, the Clarke and Park Transforms, developed by E. Clarke (1883–1959) and R. Park (1902–1994) from their work in three phase power transmission and synchronous three phase machine analysis.

Detailed description of the transformation process is beyond the scope of this book, but is contained in various technical papers and in the product literature describing the operation of μP chips available to create such transforms [28].

Essentially, the three phase stator currents (a, b and c) and the rotor velocity and position are monitored and, along with the stator and rotor parameters (resistance, inductance), are used to transform the three phase currents into two imaginary orthogonal currents, I_d and I_q simulating the currents of a wound field DC motor [29] (see Figure 3.54).

I_d represents the rotor flux and is compared to a flux reference in a PI control loop.

I_q represents the torque and is compared to a torque reference in an inner PI control loop The torque reference is obtained by comparing a speed reference to an encoder feedback signal, forming an outer velocity control loop.

The two resulting error voltages are then inverse transformed into the three phase voltage/current input to the stator.

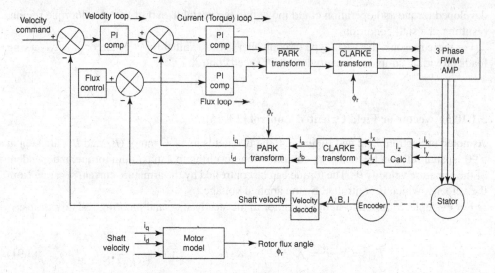

Figure 3.54   Block diagram of three phase vector control

There are two main categories of these controllers:

- Sensor type, that uses a shaft encoder to provide rotor velocity and position.
- Sensor less type, that has no encoder and uses an algorithm based on a mathematical model of the motor to estimate slip and rotor status.

One problem, in general, with the implementation of vector control is the requirement for the knowledge of certain motor parameters, some of which are not normally available for the typical "off the shelf" induction motor.

These include:

- Line frequency
- Number of poles
- Rated speed (full load)
- Rated torque (full load)
- Stator resistance and inductance
- Rotor resistance and inductance
- Mutual and leakage inductances
- Encoder resolution
- Load inertia.

Older controllers required this data to be hand entered but more recent designs contain generic data which is updated by an auto-tune program when connected to a specific motor.

3.1.10.6 The Slip Model [30]

Open loop operation of the induction motor in the stable region can be modeled by deriving an equation of the developed torque as a function of slip from the actual speed/torque curve for the motor.

Equation 3.90 for slip: $S = \frac{N_S - N}{N_S}$ can be reduced to:

$$S = 1 - \left(\frac{1}{N_S}\right)(N) \tag{3.92}$$

The following table shows this expression for six pole counts; N and N_S in rad s^{-1}.

Pole Count	Synchronous Speed (rpm)	S
2	3600	$1-0.00265N$
4	1800	$1-0.00531N$
6	1200	$1-0.00796N$
8	900	$1-0.011N$
10	720	$1-0.013N$
12	600	$1-0.016N$

Using this expression for S and the data from the speed/torque curve, a table of speed, slip and torque, from no load to stall can be created.

These data can then be entered into Excel, plotted and using the Curve Fitting Routine an equation of torque as a function of slip can be developed. This equation can then be used to simulate the motor and its response to loads.

This procedure is best understood by following an example; a 10 HP six pole motor.

Rated torque = 61 N m	Rated speed = 1160 rpm	Speed/Torque curve; Figure 3.55
Speed (rpm)	S	Torque (N m)
1200	0	0
1160	0.033	61
1140	0.05	81.3
1080	0.10	101.7
1020	0.15	114.4
960	0.20	120.0
900	0.25	122.0
840	0.30	122.0
780	0.35	120.0
720	0.40	117.0
660	0.45	114.4

(continued)

Rated torque = 61 N m	Rated speed = 1160 rpm	Speed/Torque curve; Figure 3.55
Speed (rpm)	S	Torque (N m)
600	0.50	111.0
540	0.55	106.8
480	0.60	101.7
420	0.65	96.6
360	0.70	94.0
300	0.75	89.0
240	0.80	86.4
180	0.85	86.0
120	0.90	86.4
60	0.95	89.0
0	1.0	91.5

Figure 3.56 shows the slip and torque data plotted with a fifth order curve fitted to the data. The result is the equation of torque as a function of slip:

$$T = 3911.8S^5 - 11082S^4 + 11961S^3 - 6096.4S^2 + 1387.6S + 12.863 \qquad (3.93)$$

Although the detailed calculations are not shown here, the value of torque calculated from this equation is within ±5% of the actual data.

This equation is then used to develop the simulation of the motor as shown in Figure 3.57a and b.

In Figure 3.57a the motor is accelerating to and running at no load speed.

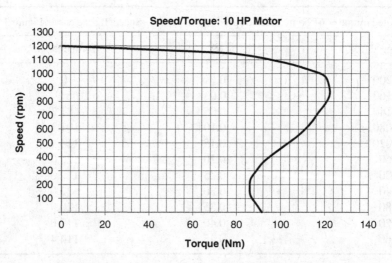

Figure 3.55   Speed/torque curve for 10 HP six pole motor

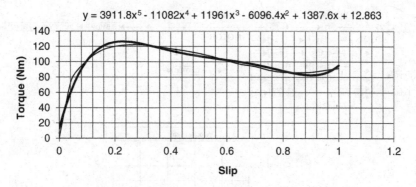

Figure 3.56 Torque/slip for 10 HP six pole motor

In Figure 3.57b after reaching no load speed, the motor is subjected to a load torque causing its speed to drop to 1170 rpm.

3.2 Gearheads

Gearheads provide two main functions in motion control systems:

- They match the relatively low speed/high torque requirement of the load to the high speed/ low torque, high efficiency operation of the motor.
- Synonymous with this, the load inertia seen at the gearhead input is the load inertia divided by the square of the gear ratio, resulting in a close match to the motor inertia (ideally 1:1), creating the highest overall system efficiency (see Section 4.5).

Although available in many configurations and technologies, gearheads using the following are those most frequently found in high performance, closed loop servo systems:

- Spur gearing
- Planetary gearing
- Hybrid gearing
- Worm gearing
- Harmonic gearing.

3.2.1 Spur Gearhead

The spur gear is the oldest design, going back to as early as 30 BC and from which all other configurations have evolved. A simple schematic for a one pass spur gear arrangement is shown in Figure 3.58.

The smaller gear, called a pinion, is usually mounted on the input shaft and the larger gear is connected to the output shaft. A single pass, as shown, is usually applicable to no more then a 10 or 15 to 1 ratio. Larger ratios are accomplished by using two or three passes, with the pinion of succeeding passes mounted on the same shaft as the larger gear of preceding passes.

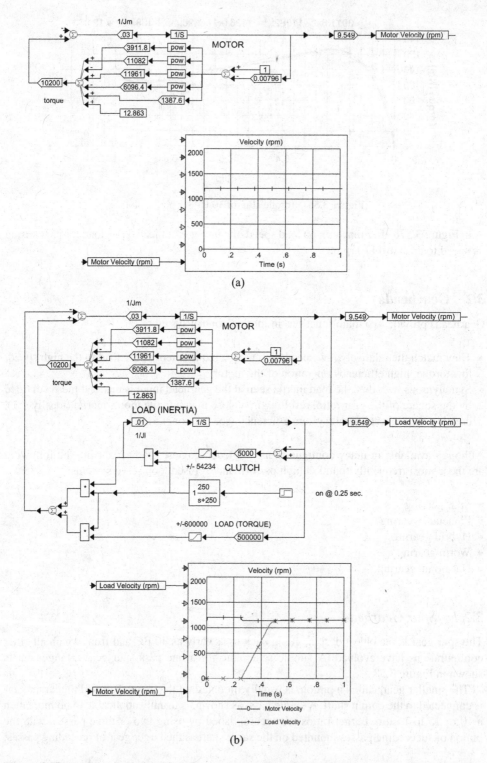

Figure 3.57   (a) Slip simulation of 10 HP six pole motor; no load. (b) Slip simulation of 10 HP motor with clutch applied load (available in full color at www.wiley.com/go/moritz)

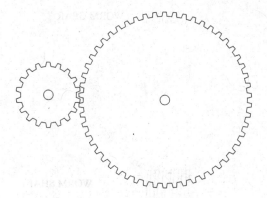

Figure 3.58 Spur gearhead schematic

One deficiency of the basic spur gear design is that in its simplest configuration, as shown, the input shaft and output shaft are not in line. This is overcome by designing multi-pass units with offset passes such that the shafts line up, resulting in units that can easily be mounted to motors, achieving a complete assembly with an inline output shaft.

Precision spur gearheads are typically available in ratios from 3:1 to 100:1, with 98% efficiency in sizes 60 to 115.

3.2.2 Planetary Gearhead

Planetary gearing is a unique arrangement in which a set of four gears concentrically arranged around an in line input/output shaft alignment is capable of providing three times the torque capacity in the same volume as a spur gear assembly (see Figure 3.59).

The three planet gears are driven by the sun gear (the input gear) and are captured by the ring gear, which is machined into the gearhead housing. The three planets are in turn mounted on a spider assembly whose center becomes the output shaft.

Planetary gearheads can be fabricated with one, two or three passes in tandem providing ratios from 3:1 to 500:1 with efficiencies of 85 to 96%, in sizes 14 to 160.

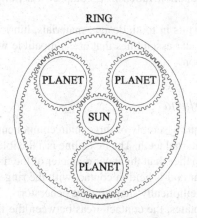

Figure 3.59 Planetary gearhead cross-section

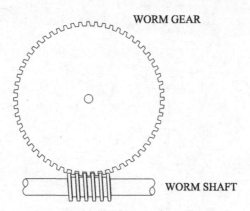

Figure 3.60   Worm gearhead schematic

3.2.3   Hybrid Gearhead

A hybrid gearhead consists of one or two input spur gear stages; with a planetary gear output stage. This arrangement can provide a high ratio design that is lower in cost than a multi-pass planetary design.

It is often used in right angle gearheads in which the input pass is a spur gear mesh followed by a right angle bevel pass that drives the output planetary gearing.

3.2.4   Worm Gearhead

A worm gear assembly consists of two components, as shown in Figure 3.60.

The input is the worm shaft, similar to a lead screw. The output is obtained from the worm gear that meshes with the worm. This arrangement is unidirectional in that the worm cannot be driven by the worm gear and actually will bind and result in a braking action if the load attempts to back drive.

Worm gearing has traditionally been used for high power, unirotational applications to obtain right angle, high reduction ratios in a compact assembly in which efficiency is of secondary concern.

Contemporary improvements in tooth design, materials, lubrication, bearings and thermal design have created worm gear assemblies that are compatible with the severe requirements of closed loop servo operation.

3.2.5   Harmonic Gearhead

The harmonic gear is a unique assembly of three main components, as shown in Figure 3.61.

The fixed ring gear has internal teeth. The flexspline is a flexible cylindrical member whose outer diameter is slightly smaller than the inner diameter of the fixed ring gear, has two fewer teeth than the fixed ring gear and is held in contact with the ring gear in only two areas 180° apart, as determined by the elliptically shaped wave generator.

As the wave generator rotates, the contact areas between the flexspline and the fixed ring gear rotate, such that when the wave generator rotates 360° the contact area only moves two

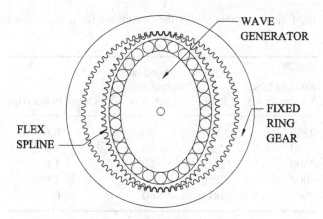

Figure 3.61   Harmonic gearhead cross-section

teeth from the starting position. With the wave generator connected to the input shaft and the flexspline connected to the output shaft, the result is a high ratio compact gearhead.

Harmonic gearheads are available with ratios of 50:1 to 160:1 with efficiencies of 65 to 85% in sizes 14 to 100.

3.2.6   Gearhead Sizing – Continuous Operation

The selection of the type and size (rating) of the gearhead for use in any application is not a straightforward process, especially in incremental motion applications involving repetitive accelerations, decelerations and variable duty cycles.

Acceleration, velocity, inertial matching, backlash, size, stiffness, maximum expected ambient temperature, and so on must all be considered and possibly modified as the selection process proceeds.

All gearheads have continuous rated nominal output torques and maximum rated nominal input velocities, typically specified at some maximum case temperature, usually 90 or 100 °C as measured in a 25 °C ambient environment.

This immediately brings to light a problem with manufacturers' data, in that they do not specify the conditions in which the thermal rating was tested. Was the gearhead thermally isolated? Was it mounted on a steel or aluminum plate? Was it thermally isolated from the driving motor? Knowing answers to these questions would help in evaluating the intended environment of the application. See Section 3.2.10 for further analysis of this situation.

Theoretically, for a continuous running application, one could evaluate the torque and velocity required at the load, settle on the motor velocity and determine the gear ratio. If the ratio is not available, select the nearest standard value and adjust the motor velocity to achieve the required result.

However, the nominal output torque and input velocity ratings, together with various ratios, do not result in a constant power capability.

For example, the following table shows the conditions for a particular size 90 planetary gearhead:

Nominal Input Velocity (rpm)	Ratio	Nominal Output Torque (N m)	Output Power (HP)
2500	3:1	56	6.6
3500	7:1	66	4.6
3500	15:1	56	1.8
4000	20:1	66	1.9
4800	100:1	60	0.4

Theoretically, if the output power were constant, the torque for the 100:1 ratio would be 1867 N m. However, the output torque capability is a function of the gear component design (tooth strength, diameter, thickness, etc.) and the bearings and, therefore, remains fairly constant regardless of velocity.

The internal temperature is caused by three main effects, all generated by the velocity; the churning of the lubricant, the shaft oil seal and bearing friction. The internal temperature must be limited to 120 °C, the maximum temperature rating of the synthetic hydrocarbon lubricant typically used in gearheads.

3.2.7 Gearhead Sizing – Intermittent Operation

Gearhead manufacturers have created various procedures to be used to correlate the torques experienced during intermittent operations, such as trapezoidal velocity profiles, with the nominal constant ratings, to enable gearhead selection.

In general, the first step is to calculate a torque variously designated as:

$$T_{mean}, \ T_{average} \ \text{or} \ T_{equivalent}$$

Since the following equation is not a true averaging expression, $T_{equivalent}$ will be used.

$$T_{equivalent} = \sqrt[x]{\frac{N_1 t_1 T_1^x + N_2 t_2 T_2^x + N_3 t_3 T_3^x + \ldots N_n t_n T_n^x}{N_1 T_1 + N_2 T_2 + N_3 T_3 + \ldots N_n T_n}} \qquad (3.94)$$

where:

N_n and T_n = the average velocity and torque respectively during period t_n

x = a factor dictated by the gearhead manufacturer; a survey of five manufacturers results in values for x of:

3, 3/10 and 8.7

For a typical trapezoidal velocity profile:

$$t_1 = t_{acc}, \ t_2 = t_{run}, \ t_3 = t_{dec}, \ t_4 = t_{stop}$$

$$N_2 = \text{run velocity}, \ N_1 = N_3 = N_2/2$$

$$T_1 = T_{acc}, \ T_2 = T_{run}, \ T_3 = T_{dec}, \ T_4 = T_{stop}$$

Example
Assume a trapezoidal profile in which:

$$N_2 = 400 \text{ rpm}, \ N_1 = N_3 = 200 \text{ rpm}$$

$$t_1 = 3 \text{ s}, \ t_2 = 20 \text{ s}, \ t_3 = 3 \text{ s}, \ t_4 = 0$$

$$T_1 = 60 \text{ Nm}, \ T_2 = 15 \text{ Nm}, \ T_3 = 60 \text{ Nm}, \ T_4 = 0$$

Then for the three different x values;

$$T_{equivalent} = 31.5 \text{ Nm for } x = 3$$

$$= 33.2 \text{ Nm for } x = 10/3$$

$$= 47.5 \text{ Nm for } x = 8.7$$

The next step is to calculate what is designated as the "mean" or "average" velocity. For this characteristic, all the manufacturers agree on the following formula:

$$N_{mean} = \frac{N_1 t_1 + N_2 t_2 + N_3 t_3 + \cdots\cdots N_n t_n}{t_1 + t_2 + t_3 + \cdots\cdots s t_n} \tag{3.95}$$

For this example: $N_{mean} = 354$ rpm

From this point, each manufacturer outlines a series of steps to be used that will result in selecting the appropriate gearhead. The following is a general summary of the selection procedure:

- Choose a gearhead whose T_{rated}, T_{cont}, and so on. is larger than $T_{equivalent}$
- Determine that $T_{peak\text{-}repeated}$, $T_{acc\ rated}$, and so on. is larger than the greater of T_{acc} or T_{dec}
- Calculate $\frac{N_{max\ rated}}{N_{run}}$ to determine maximum possible ratio
- Select a standard ratio "R" that is below this value
- Calculate the input mean velocity as $R \times N_{mean}$
- Calculate the input peak velocity as $R \times N_{run}$
- If either of these exceed the gearhead velocity ratings, select a lower "R"
- Determine that $T_{emergency}$, $T_{momentary}$, and so on. is compatible with possible emergency conditions.

At this point, the axial load, radial load, backlash and stiffness ratings of the selected gearhead should be examined for compatibility with the system requirements.

3.2.8   Axial and Radial Load

The axial and radial load ratings of a gearhead are essentially the ratings of the output bearing structure. As such, the ratings will vary from manufacturer to manufacturer depending on the type, size and class of bearing plus the bearing support structure each supplier has designed into his product line. These ratings typically supplied in graphical, tabular or formulaic form.

Radial load ratings are usually specified at the center of the shaft at 100 rpm, with plots or equations that can be used to determine the rating at different distances and speeds.

In addition, these ratings are based on specific service life factors, typically 10 000, 20 000 or 30 000 hours at 100 rpm. Applications involving higher velocities can reduce the service life significantly.

Unless specified otherwise, the axial and radial load ratings are independent, that is, they are each specified assuming the other is at zero. If the application involves simultaneous axial and radial loads, the effect of this on the ratings and the service life should be reviewed with the manufacturer.

3.2.9   Backlash and Stiffness

These two terms have been defined many different ways and often in ways made to make a product appear to be better than it actually is. For example, some manufacturers claim to have a gearhead with "no backlash".

The fact is, all gearheads, regardless of type (spur, planetary, worm, harmonic, etc.) have backlash for two main reasons:

- There will be clearance (a gap) created by the fact that the space between one pair of teeth on one gear will be larger than the width of the tooth on the mating gear simply due to manufacturing tolerances, as shown in Figure 3.62.
- Even if the mating parts were "perfect" and no gap existed, a gap would have to be created simply to provide space for the lubricant that is necessary for proper, low friction operation of the gears.

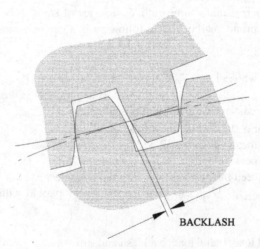

BACKLASH

Figure 3.62   Backlash

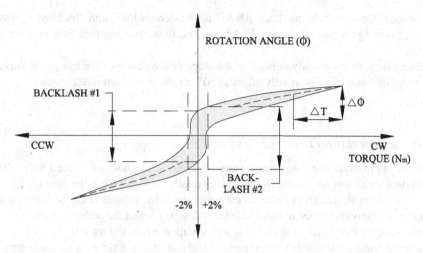

Figure 3.63 Backlash and stiffness diagram

No backlash would mean metal to metal contact, resulting in high losses and metal wear and fatigue. The claim of no backlash by a manufacturer should be closely examined and questioned as to his interpretation of the term and how he measures it.

The question is not one of does a gearhead have or does not have backlash, but how to measure backlash and interpret the results of the measurement.

Backlash, and stiffness which will be covered further on, is measured by clamping the gearhead in a solid fixture, fixing the input shaft so it cannot rotate and then rotating the output shaft.

In a perfect, theoretical, frictionless gearhead the output shaft could be slowly rotated in one direction until opposition is felt, then the direction reversed until opposition is again felt. The total backlash is then the angle through which the shaft is rotated under the condition of no torque.

However, all gearheads have friction, meaning that in order to rotate the output shaft, a certain amount of torque must be applied and this makes it difficult to determine the point at which contact is made and the rotation should be reversed.

Since the point of contact is difficult to determine, many manufacturers have settled on a procedure that involves applying 2% of the rated torque in both direction and defining the resulting angular displacement as the backlash.

Also, instead of trying to measure only backlash, both backlash and stiffness can be measured by applying torque, starting at zero and increasing it to the maximum rated value, recording angular displacement versus torque as the torque is increased to maximum in a CW direction, decreased back to zero, then increased to maximum in a CCW direction and then finally decreased back to zero. The result will be the angle versus torque hysteresis curve as shown in Figure 3.63.

From this curve, the "ideal" or "exact" backlash can be determined, shown as "backlash #1" where the plot crosses the rotation angle axis for zero torque.

The testable or more easily defined backlash, shown as "backlash #2", is shown as the rotation angle values where the curve intersects the +2% and −2% torque values.

Some manufacturers refer to "backlash #1" as "hysteresis loss" and "backlash #2" as "lost motion", avoiding the use of the word backlash and then claiming that their product has no backlash.

Stiffness is more universally defined as the slope of the center of the hysteresis curve, $\frac{\Delta\phi}{\Delta T}$, as arc min/Nm, where ΔT is usually taken as 50% of the maximum rated torque.

3.2.10   Temperature/Thermal Resistance

Except for stating that the case temperature of a gearhead must not exceed 90 or 100 °C, thermal data are essentially nonexistent. This is totally different from the case for motors, in which thermal resistance from stator to case and stator to ambient under specific mounting conditions, is always provided to help the system designer select the proper conditions to avoid thermally stressing the insulation and magnetic components of the motor.

In order to bring some clarity to gearhead thermal conditions, a series of tests was performed on planetary gearheads as outlined in the following data for one of the tests.

For the test, the gearhead was mounted on a thermally isolated $25.4 \times 25.4 \times 1.27$ cm^3 aluminum plate in a vertical orientation. The gearhead was driven by a DC motor through a thermally isolated torque meter. Thermocouples were used to measure the internal and case temperatures.

The gearhead was operated at no load to establish the no load losses and thermal resistance as a function of input velocity.

Gearhead: Size 90 planetary, single pass, 10:1 ratio, efficiency = 97%

Input Vel. (rpm)	Torque (N m)	Power (W)	t(int) (°C)	t(case) (°C)	t(amb) (°C)	int-case	Thermal R (°C W^{-1}/ watt) int-amb
250	0.148	3.87	25.38	24.17	22.03	0.313	0.866
500	0.192	10.1	29.53	27.89	22.52	0.162	0.694
750	0.219	17.3	32.77	30.63	22.45	0.124	0.597
1000	0.245	25.7	36.96	34.11	22.21	0.111	0.574
1250	0.283	37.1	39.55	36.36	22.18	0.086	0.468
1500	0.298	47.0	42.65	39.13	22.38	0.075	0.431
1750	0.313	57.4	44.29	40.92	22.26	0.059	0.384
2000	0.345	72.3	46.39	43.22	22.38	0.044	0.332
2500*	0.384*	100.7*					0.313*
3000*	0.434*	136.5*					0.269*
3500*	0.484*	177.7*					0.233*
4000*	0.534*	223.9*	73.90	72.80	23.30	0.0047	0.226*

*These data were derived from the 250–2000 rpm test data as analyzed and described in the following paragraphs.

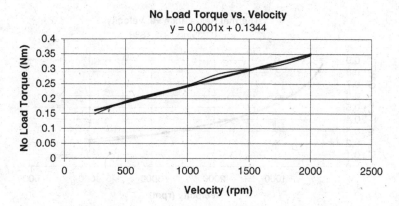

Figure 3.64 No load torque/velocity; 250 to 2000 rpm

Figure 3.64 is a plot of torque versus velocity (for 250–2000 rpm), showing a nominal linear relationship.

A curve fitting routine was then implemented, shown in Figure 3.64, and the resulting equation,

$$y = 0.0001x + 0.1344 \tag{3.96}$$

was used to calculate the additional torque* and power* data points for 2500, 3000, 3500 and 4000 rpm shown in the table and plotted in Figure 3.65.

Next, a plot of the internal-to-ambient thermal resistance versus velocity was prepared, as shown in Figure 3.66.

A curve fitting routine was again implemented, shown in Figure 3.66, and the resulting equation:

$$y = -0.2391 \ln(x) + 2.183 \tag{3.97}$$

was used to calculate the additional thermal resistance data points* for 2500, 3000 and 3500 rpm shown in the table.

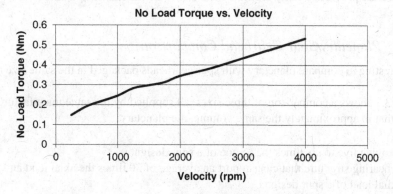

Figure 3.65 No load torque/velocity; 250 to 4000 rpm

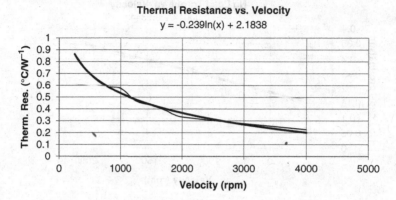

Figure 3.66   Internal to ambient thermal resistance/velocity

This gearhead has a maximum continuous rating of 60 N m at 3500 rpm input velocity and an efficiency of 97%. This is an output of 2201 W, a load input of 2269 W and a loss of 68 W.

$$\text{Total dissipation} = 68 + 177.7 = 246 \text{ W} \tag{3.98}$$

$$\text{Internal temperature} = 246 \times 0.233 + 25 = 82 \text{ }^\circ\text{C}, \tag{3.99}$$

which is within 10% of the maximum 90°C rating.

These test data and calculations show some interesting characteristics about planetary gearheads:

- The no load losses are linearly proportional to velocity
- The thermal resistance is not constant, and is inversely proportional to velocity in a logarithmic relation.
- Over the complete velocity range, the case temperature is no more than three degrees lower than the internal temperature. Therefore, for any mounting arrangement, the case temperature provides a fairly accurate measure of the internal temperature.

3.2.11   Planetary/Spur Gearhead Comparison

It is interesting to compare planetary with spur gearheads packaged in the same size (volume) enclosure.

Table 3.1 shows a comparison of three sizes, all supplied by the same manufacturer.

Note that in approximately the same volume, the planetary:

- Can provide five to six times the torque of a spur design.
- Has a bearing structure that can support an average of 30 times the axial load and 15 times the radial load of a spur design
- Has an average of one fifth the backlash of a spur design.

Table 3.1 Planetery versus spur gearhead comparison

Type	Volume (cm³)	Maximum Torque (N m)	Nominal Speed (rpm)	Maximum Power (HP)	Axial Load (N)	Radial Load (N)	Backlash (arc min)
60 Spur	195	6	4000	0.34	45	90	30
60 Planetary	230	37	5200	3.8	2100	1650	6
90 Spur	524	20	4000	1.4	135	350	25
90 Planetary	563	110	4800	8.9	3600	4800	6
115 Spur	1089	40	4000	2.6	265	890	25
115 Planetary	1193	230	4200	16.1	6800	7500	4

3.3 Leadscrews and Ballscrews

These components basically operate in the same manner. A nut in contact with an externally threaded rod, the screw, moves laterally as the screw is rotated.

The major difference between a lead screw and a ball screw is the nature of the nut. In a lead screw an internally threaded nut mates with the screw. Contact between the two members consists of sliding friction. In a ball screw, the sliding friction of the lead screw nut is replaced with the rolling friction of ball bearings which circulate around the screw threads as the screw rotates. See Figures 3.67 and 3.68.

In the majority of applications, the assembly is used to convert rotary motion into linear motion by driving the screw with a motor (DC, AC or stepper) and having the load, connected to the nut, move laterally.

There are also assemblies in which the nut is mounted internal to the rotor of a motor and the screw moves laterally as the rotor/nut rotates.

In addition, due to its low friction and high efficiency, it is possible to move the nut of a ball screw assembly laterally, causing the screw to rotate in a fixed mount.

The following specifications represent a summary of data compiled from a number of manufacturers.

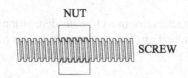

Figure 3.67 Lead screw cross-section

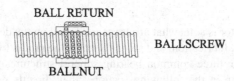

Figure 3.68 Ball screw cross-section

3.3.1 Leadscrew Specifications

- Diameter: 0.318 to 7.62 cm
- Length: Not specified; requires discussion about application with vendor
- Screw materials: Low carbon steel for diameter less than 2.54 cm.

  Medium carbon steel for diameter greater than 2.54 cm.

  Stainless steel, type 304

  A6 tool steel.

- Nut naterials: Steel

  Bronze – for high loads – requires lubrication

  Plastic – for low loads – no lubrication required.

- Efficiency: 10 to 40%, depending on lead angle, nut material and friction coefficient. Efficiency as high as 87% available with special nut finish.
- Weight: Typically given as "per foot" value
- Inertia: Not given/requires user to calculate
- Speed: Less than 100 rpm typical. Up to 300 rpm for light loads.

3.3.2 Ball Screw Specifications

- Diameter: 1 to 4 cm
- Length: Not specified; requires discussion about application with vendor
- Screw material: High carbon or alloy steel – case hardened
- Nut Materials: carbon or alloy steel – chrome steel balls
- Efficiency: 90% or higher
- Weight: Typically given as "per foot" value
- Inertia: Not given/requires user to calculate
- Speed: Up to 5000 rpm.

The two most important parameters that must be considered for a ball screw/lead screw design are critical speed and column loading. They are both functions of:

- Screw diameter
- Unsupported screw length
- Method of bearing support at each end
- Safety factor.

The choice of safety factor is somewhat arbitrary and based on the designer's experience, the duty cycle, environment, lubrication, and so on to which the product will be exposed.

Figure 3.69 shows the three common bearing support structures used in ball screw/lead screw designs. As shown in the following sections, they directly determine the maximum permissible velocity and load capacity for any particular diameter and screw length.

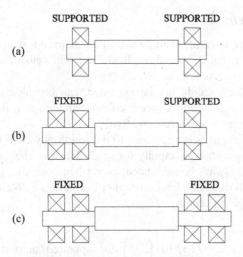

Figure 3.69   Screw bearing structures

3.3.3   Critical Speed

Any rotating shaft supported at its extreme ends will have a natural frequency of vibration determined by the shaft length and diameter. In designing a ball screw/lead screw assembly it is important to calculate the critical speed and restrict the maximum operating velocity to fall well below the critical speed to avoid vibration and possible component failure.

For the three basic bearing supports shown in Figure 3.69, the following formulas allow the calculation of the associated critical speeds:

$$N_{SS} = \frac{12.1 \times 10^6 \times d}{l^2 \times SF} \text{ for supported/supported} \qquad (3.100)$$

$$N_{FS} = \frac{18.7 \times 10^6 \times d}{l^2 \times SF} \text{ for fixed/supported} \qquad (3.101)$$

$$N_{FF} = \frac{27.2 \times 10^6 \times d}{l^2 \times SF} \text{ for fixed/fixed} \qquad (3.102)$$

where N is the velocity in rpm, d the minor diameter in cm, l the length in cm, SF the safety factor, 1.25 to 2.5 (see previous note).

These formulas demonstrate that there are a number of trade-offs to consider when developing a screw system design, all of which have a direct impact on the system cost.

Example
For a fixed length, the critical speed can be increased by increasing the diameter which would increase cost. Instead, the design could be changed to a more fixed bearing structure which would also increase cost, possibly more than the increased cost of a larger screw. However, increasing the screw diameter will also increase the screw inertia and, depending on the screw inertia to load inertia ratio, would affect peak acceleration power required from the motor and drive electronics.

3.3.4   Column Strength

Screws which are loaded in compression when an unbalanced force is acting along the screw axis or in a vertical orientation where the load itself will cause compressive loading, can experience screw deformation.

Maximum loads, which should not be exceeded, can be calculated with the following formulas. As in the case for critical speed, column strength is a function of minor screw diameter, unsupported length and screw end fixity.

Note that column strength is proportional to the fourth power of the minor diameter. As such, the column strength changes rapidly for small diameter changes. Because of this, the column strength should always be calculated, even when designing a system with relatively light loads using a small diameter, long screw. For example, a 20% change in screw diameter results in a 100% change in column strength.

$$P_{SS} = 9.71 \times 10^5 \left(\frac{d^4}{l^2} \right) \text{ for supported/supported} \tag{3.103}$$

$$P_{FS} = 19.4 \times 10^5 \left(\frac{d^4}{l^2} \right) \text{ for fixed/supported} \tag{3.104}$$

$$P_{FF} = 38.8 \times 10^5 \left(\frac{d^4}{l^2} \right) \text{ for fixed/fixed} \tag{3.105}$$

where P is the column strength in kg, d the minor diameter in cm, and l the length in cm.

3.3.4.1   Example 3

A 23 000 g load is to accelerate to 30 cm s^{-1} in 0.2 s and travel 90 cm.

$$\text{Slew velocity} = N_S = \frac{30 \times 60}{L} \text{ rpm } (L = \text{screw lead in cm/rev}) \tag{3.106}$$

Using Equations 3.100–3.102 and 3.106 the following table can be prepared for a sampling of screw leads and sizes.

	L	N_S	d	N_{SS}	N_{FS}	N_{FF}
1.	0.3175	5670	0.762	760	1178	1700
2.	0.508	3540	1.22	1215	1883	2720
3.	0.635	2830	2.08	2070	3208	4640
4.	0.846	2130	1.91	1900	2945	4260
5.	1.27	1420	2.08	2070	3208	4640
6.	2.54	710	2.86	2850	4418	6380

d = minor diameter (cm); N_{XX} = rpm (safety factor = 1.5).

Examining these data shows the following:

- Items 1 and 2 cannot be considered since the small leads create velocities that exceed those available, regardless of the fixity.
- Item 6 can meet the velocity requirement but is the largest screw, will have the highest inertia and would most likely be the most expensive.
- Items 3, 4 and 5 can meet the requirements with a N_{FS} fixity and item 5 can also meet the requirement with a N_{SS} fixity. They all have the same nominal minor diameter (2 cm) and will have the same column strength and inertia.

Column strength:

$$P_{SS} = 9.71 \times 10^5 \left(\frac{2^4}{90^2}\right) = 1920 \text{ kg} \quad P_{FS} = 19.4 \times 10^5 \left(\frac{2^4}{90^2}\right) = 3830 \text{ kg}$$

For this application column strength is not a factor
Screw inertia:

Material $= Fe; \rho = 7.86 \text{ g cm}^{-3}$

$$J_{SCREW} = \frac{\pi \rho l r^4}{2g} = \frac{\pi \times 7.86 \times 90 \times 1.27^4}{2 \times 980.6} = 2.95 \text{ g cm s}^2 \quad l = 90 \text{ cm}$$

$d = 2.54 \text{ cm}$

Reflected load inertia:

$$J_{REFL} = \frac{W/g}{(2\pi P)^2} \tag{3.107}$$

The following table can be prepared for items 3, 4 and 5.

	θ' (rad s^{-1})	θ'' (rad s^{-2})	J_{REFL} (g cm s^2)	J_{TOTAL} (g cm s^2)	T_{ACC} (g cm)	P (W)
3.	296	1480	0.24	3.19	5246	153
4.	223	1115	0.425	3.38	4187	91
5.	149	745	0.958	3.91	3235	47

T_{ACC} calculated with an assumed efficiency of 90%:
$P =$ peak power at termination of acceleration (at 0.2 s).

Note that even though the reflected inertia increases by a factor of four times between item 3 and item 5, the total inertia only increases by 23% due to the predominance of the screw inertia.

For this application, item 5 is the screw to use based on lowest velocity, torque and power.

3.3.4.2 Example 4

Same as example 1, except the load is increased to 230 000 g (ten times greater).

The summary table for items 3, 4 and 5 now becomes:

	θ' (rad s^{-1})	θ'' (rad s^{-2})	J_{RFFL} (g cm s^2)	J_{TOTAL} (g cm s^2)	T_{ACC} (g cm)	P (W)
3.	296	1480	2.4	5.35	8798	256
4.	223	1115	4.25	7.2	8920	195
5.	149	745	9.58	12.53	10372	151

Now item 5 has the highest torque although it still has the lowest velocity and power. In this case, item 4 would be the best choice with the torque varying only 2 to 1 for a 10 to 1 variation of the load.

These examples show that ball screw/lead screw selection should always involve analyzing the complete range of expected load values and evaluating a number of screw sizes and leads to determine the best performance/economic combination.

3.3.5 Starts, Pitch, Lead

These three terms define the design of the thread in a /ball screw/lead screw:

- Starts defines the number of independent threads in a screw. Typically screws have one start, that is, there is a single thread running along the length of the screw.

  If a screw has two starts, then there are two independent threads alternately running along the length of the screw. In rare cases there can be as many as 20 starts, but two, three or four are typical.
- Pitch (screw terminology) is the axial distance between two adjacent threads, regardless of the number of starts. For a single start design, pitch is the distance between adjacent points on the single thread. For a two start design, pitch is the distance from one point on the first thread to the adjacent point on the second thread. Pitch (screw terminology) has the units of distance (mm, cm etc.)
- Lead is the distance the nut will move longitudinally for a single rotation of the screw. Lead has the units of distance per rotation (mm/rev, cm/rev etc.)

  Pitch and lead are functions of start as follows:

$$lead = pitch \times number\ of\ starts$$

Therefore: For the typical one start thread, lead = pitch

For a two start thread, lead = 2 pitch, that is, the nut will travel twice the pitch distance for a single rotation of the screw

For "n" starts, the nut will travel "n" times the pitch distance.

- Pitch (servo terminology).

  In motion control and servo design, pitch is defined as the reciprocal of lead and has the units of rotation per distance (rev/mm, rev/cm etc.).

  This is the pitch used in various formulas in this book.

3.3.6 Encoder/Lead

When a ball screw/lead screw is used with an encoder, the screw lead, encoder count and maximum velocity must all be considered. The smaller the lead the higher the rotational velocity must be to achieve a higher translational velocity. But the rotational velocity is limited by the encoder count (maximum frequency response) of the encoder.

A small lead together with a high encoder count will create a high resolution, but may be the factor, rather than the critical speed, that determines the maximum translational velocity.

Both the encoder count and the screw accuracy will determine the overall translational accuracy and repeatability of the system.

3.3.7 Accuracy

See Section 4.1.3.

3.3.8 Backdrive – Self-Locking

With the proper lead angle (typically less than 10°) lead screws can be self-locking, that is, a longitudinal force applied to the nut will not cause the screw to rotate (back drive).

This feature can be used to avoid the need for a brake in vertical applications.

Ball screws, due to their high efficiency, which is independent of lead angle, do not exhibit this characteristic and must use a brake or have the closed loop servo maintain zero velocity.

3.3.9 Assemblies

Ball screw assemblies typically are available in single or multi-axis configurations. They come with standard motor mounts and/or couplings to be compatible with a servo motor or stepper motor of the user's choice. In addition, linear stages are also available with a brushless motor having its rotor plus encoder mounted directly on a screw shaft extension, providing high compliance/wide bandwidth operation.

All the basic mechanical design has been done by the supplier and the user has to do the final engineering, that is, select the assembly based on distance, velocity, accuracy, size, load weight and mounting method.

Standard units are available in a range of sizes, but special requirements, especially length, are available. For a one-of-a-kind design, much time can be saved by choosing an appropriate complete assembly

Typical specs:	Stroke length: up to 2 m
Speed:	1.3 m s^{-1} max
Repeatability:	±0.01 mm, 0.02 mm, 5 μm
Thrust force:	163 kg max.
Accuracy:	15 μm, ±0.1 mm/300 mm
Max Acc.:	124 m s^{-2}
Lead:	2.5, 4, 5, 10, 16, 20, 30 mm/rev
Load:	150 kg max.
Inertia:	Specified with respect to the drive motor shaft

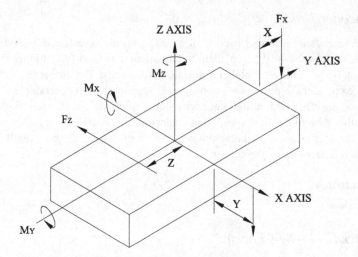

Figure 3.70   Slide loading diagram

When selecting a linear slide, an evaluation of the load weight and location with respect to the center of the carriage is important. See Figure 3.70.

Both load forces and torques must not exceed suppliers' design specifications. Note that offset loads and resulting torques must be limited to specified values to assure bearing life. These specifications are with respect to dynamic (acc/dec) values rather than static loads and must therefore be carefully calculated.

Also in X–Y and X–Y–Z assemblies, offset loads, for example a horizontal Y-axis mounted on an X-axis will subject the X-axis to a large torque load when the Y-axis (bearing the load) is fully extended.

3.4   Belt and Pulley

A belt and pulley assembly is one of the oldest methods to create linear motion from a rotary source.

The original configuration was what is now called a "power drive" in which a belt connects two pulleys, one of which is connected to the source of power (a steam engine, a diesel engine, a motor, etc.) and the other is connected to the load (a pump, a compressor, a machine spindle, etc.). In this arrangement the power pulley is usually small compared to the load pulley to achieve high torque at low velocity at the load.

The side of the belt being pulled by the power pulley is in tight tension (T_t) while the side of the belt returning from the load pulley is in slack tension (T_s).

The effective force supplying power to the load is:

$$T_e = T_t - T_s \tag{3.108}$$

There is no load on the belt or the supply power source other than the belted load and operation is usually continuous.

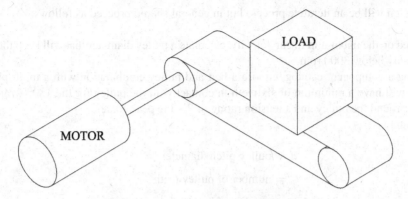

Figure 3.71   Belt and pulley schematic

The more recent configuration is the "positioning drive" in which the belt connects two equally sized pulleys with a load attached to the belt, as shown in Figure 3.71.

The use of the belt and pulley instead of a ball screw system is mainly based on the ability of the belt and pulley to achieve higher velocity over longer distances than the ball screw, as shown later in a comparison chart.

One pulley is connected to a power source (a servo motor) and the other is simply an idler pulley. Operation is intermittent with the load being driven to various locations between the end limits of the assembly according to a motion profile. The load is typically connected to a ball bearing supported plate.

In this case, the tension (force) that must be supplied by the motor is the sum of a number of components:

- Acceleration/deceleration force for the primary load; known per the system requirements
- Bearing friction load force; estimated or available from bearing supplier
- External load force, bi- or unidirectional; known per system requirements
- Gravity force if assembly is on an incline; known per system requirements
- Force to accelerate/decelerate the belt; calculable once design has initiated
- Force to accelerate/decelerate the pulleys: calculable once design has initiated.

See Section 4.9.5 for a detailed listing and a sample application.

Once the various forces are known, along with the velocity profile defining maximum velocity, timing and RMS values, the design can proceed in one of two directions:

- A catalog/vendor search can be started to find a belt driven assembly that will meet the system requirements. In this case the system designer has no control over the internal design of the assembly, similar to choosing a gearhead.
- A design can be started to use available belt and pulley parts from component vendors to fabricate the motion system as part of the overall equipment, requiring some knowledge of the detailed design requirements of belt and pulley assemblies.

The design will be an iterative process but in general should proceed as follows:

1. Based on the maximum linear velocity, calculate a pulley diameter that will keep the motor velocity below 3000 rpm.
2. Using a component catalog, choose a belt and pulley combination with a tooth pitch (p) that will have a minimum of six teeth in contact with the pulley for the 180° wrap of the belt around the pulley and a tension rating of 2× the peak force.
3. Define:

$$d = \text{pulley pitch diameter}$$

$$t_p = \text{number of pulley teeth}$$

$$t_b = \text{number of belt teeth}$$

$$L = \text{length of belt}$$

$$C = \text{distance between center of pulleys}$$

4. Calculate the following:

$$d = \frac{p \times t_p}{\pi} \tag{3.109}$$

$$L = p \times t_b \tag{3.110}$$

$$L = 2C + \pi d \tag{3.111}$$

Repeat the calculations as necessary making changes until they all agree, C being the most flexible since the other values are determined by catalog information.

5. Contact the component supplier during this process for guidance in choosing the correct components.

3.4.1 Belt

The older power systems typically use flat or V belts which depend on friction for their operation. They are subject to slipping and as such generate heat.

For precision positional applications, a toothed timing belt, assumed in the previous calculations, is the proper belt.

3.4.2 Guidance/Alignment

For proper tracking, especially at the high velocities experienced in precision motion systems, timing belts require:

- Flanged pulleys for proper tracking
- The axes of the two pulleys should be parallel to minimize unequal loading across the belt width and wear against the flanges.

3.4.3 Belt and Pulley versus Ball Screw

This listing is a comparison between two sizes of belt and pulley versus ball screw assemblies from the same manufacturer.

The belt and pulley has longer stroke and faster velocity than the ball screw.

The ball screw has lower mechanism repeatability than the belt and pulley. However, in a closed loop configuration using a linear encoder, better performance is possible.

	Belt and Pulley		Ball Screw	
Size	18	63	18	63
Stroke (mm)	1000	4500	500	2000
Max speed (m s^{-1})	2	5	0.2	1.2
Max speed (Rpm)	2300	1700	3000	2400
Repeatability (mm)	±0.08	±0.1	±0.02	±0.02
Max force (N)	60	1500	140	1600

3.5 Rack and Pinion

A rack and pinion mechanism, as shown in Figure 3.72, directly converts rotary motion into linear motion without any intermediary component.

It can be thought of as a gearhead in which one of the gears has an infinite diameter.

As shown in Figure 3.72, the load is mounted on or with the rack and will move in a linear fashion when the motor, stationary, rotates. Position control can be achieved with a rotary encoder mounted on the motor or with a linear encoder mounted on the load subsystem, similar to the arrangement in a ball screw drive.

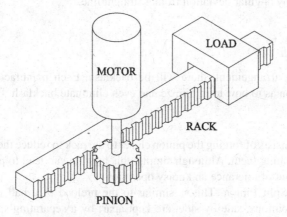

Figure 3.72 Rack and pinion schematic

One advantage the rack and pinion has over the ball screw is that it can be used over much longer length. The ball screw has length limitations due to its whipping characteristic, as described in Section 3.3.

The rack components typically are available in standard lengths (1 m for example) and can be butted together to create extremely long travel capability, only limited by the designer's requirements.

Note, however, that the space required for the mechanism must be twice the length of the travel of the rack since it moves back and forth under the pinion.

This can be avoided by mounting the motor on the load and having it move the load/motor assembly along a stationary rack. The motor essentially pulls itself along the rack. One negative aspect of this configuration is that the motor connecting cable must be designed as a slack loop capable of moving the full travel of the load motion.

3.5.1   Design Highlights

Designing an assembly to operate with a rack and pinion involves close cooperation between the machine designer and the rack and pinion supplier. Unlike designing a system which uses a gearhead, in which the internal components are not under the designer's control, a rack and pinion will eventually be assembled by the machine builder and will have to be designed to meet certain dimensional requirements to provide proper operation. Essentially, the machine designer will be involved, with the supplier, in the design of a "gearhead". As such, he must consider a number of requirements such as:

- The size and type of gear teeth must be determined by the "force" rating of the components; namely the maximum linear force expected, as determined by the maximum acceleration and deceleration.
- The machine must be designed to provide lubrication of a type specified by the manufacturer, for the full length of travel.
- If multiple racks are to be used, they must be butted properly to maintain accuracy, they must be mounted on a specified flat surface supported by a flat bearing structure and they must not have any angular deviation from a straight line.

3.5.2   Backlash

As in any gearing arrangement, there will be backlash. Each manufacturer has their own method and opinion as to how to minimize and even eliminate backlash. Essentially, they are the following:

- Preload. This consists of forcing the pinion closer to the rack to reduce the nominal clearance between the meshing teeth. Although simple, this method can lead to premature wear and possibly cause out of tolerance and noisy operation.
- Spring Loaded Split Pinion. This is similar to the preload of a ball nut in a ball screw assembly. Two pinions, side by side, are kept apart by a separating spring so they make contact on adjacent teeth, essentially filling the entire tooth gap. The tension provided by

the spring must be high enough so there is no motion between the two pinions under the highest force to be encountered in operation.

- Servo Controlled Split Pinion. Two pinions are separately controlled, one operating as the master in response to normal commands and the second creating a dynamic control of backlash. At zero velocity the slave creates a force in opposition to the master. When accelerating or running at constant velocity the slave changes from opposition to aiding the master, thus eliminating the force reduction experienced in the spring-loaded design.

3.5.3   Dynamics

The dynamic evaluation should consist of first determining the linear values of load mass (weight), slew velocity, acceleration, deceleration, load forces, velocity profile times and duty cycle and then translate these to the pinion/motor. Basic equations to perform this are covered in Section 4.9.6.

Two items of interest are:

- Derivation of load mass translated as equivalent inertia at the motor shaft

$$\text{At the rack: } F = mS'' \tag{3.112}$$

$$\text{At the motor shaft: } T = F\left(\frac{D}{2}\right) = mS''\left(\frac{D}{2}\right) \tag{3.113}$$

$$\text{Since: } S'' = \theta''\left(\frac{D}{2}\right) \tag{3.114}$$

$$T = m\theta''\left(\frac{D}{2}\right)^2 = \left(\frac{W}{g}\right)\left(\frac{D}{2}\right)^2\theta'' \tag{3.115}$$

$$\text{Equivalent inertia} = \left(\frac{W}{g}\right)\left(\frac{D}{2}\right)^2 \tag{3.116}$$

where F = force, T = torque, S'' = linear acceleration, m = load mass, D = pinion diameter, θ' = rotary acceleration, W = load weight, and g = acceleration due to gravity.
- Motor Velocity/Gearhead Requirement.

$$\text{The motor velocity is: } \theta' = \left(\frac{S'}{r}\right)\left(\frac{60}{2\pi}\right) \text{ rpm} \tag{3.117}$$

where r = pinion radius, S' = linear velocity, and θ' = rotary velocity.
- This expression can be used as a simple approximation to determine the need for a gearhead between the motor and the pinion.
- If θ' is in the high hundreds, or above 1000 rpm, a gearhead is most likely not needed. If θ' is in the range of 100 to 500 rpm, which is usually the case for heavy low velocity loads, a gearhead should be considered for a more efficient design and possibly a smaller motor than with a direct drive.

3.6   Clutches and Brakes

Clutches and brakes are conceptually fairly simple devices that either connect a load to a torque source to create motion (clutch) or to ground to stop motion (brake).

They are also combined into a single unit (clutch/brake) to provide intermittent run/stop operation.

Although run/stop action can be provided by a closed loop position servo, high cycling rates involve continuous acceleration and deceleration of the servo motor, leading to relatively high dissipation. If positioning accuracy will allow, especially in fixed motion profiles, a clutch/brake approach can lead to lower cost and lower dissipation than the servo approach. However, since modern controllers allow a servo to create controlled acceleration, deceleration and jerk, precision and variable motion is still the preferred method for incremental motion system design.

Many technologies are used in the design of clutches and brakes. One of the most popular, the subject of this section, is the electromechanical type, in which action is initiated by either applying or removing electrical power to a control coil.

3.6.1   Clutch/Brake Types

The four basic methods by which the connection between the input and output of the clutch or brake is accomplished are:

- Friction
- Magnetic particle
- Eddy current
- Hysteresis.

3.6.1.1   Friction

A friction unit contains two mating surfaces, one steel and the other an organic material. When forced together, the two surfaces "lock" together and transmit torque from the input to the output (clutch) or from the output to ground (brake). See Figure 3.73

(a) Spring/electromagnet type

In a clutch, a spring keeps the two surfaces separated. When a control coil is excited, an electromagnetic field opposes the spring force and causes the two surfaces to connect.

In a brake, the same action creates "power on braking ", typically used together with a clutch for rapid start/stop applications. Reversing the action, in which exciting the coil keeps the two surfaces separated, and turning the coil off allows the spring to force the two surfaces together creating "power off braking" for the typical fail safe application.

(b) Permanent magnet/electromagnet.

This type is used for brake designs. A permanent magnet replaces the spring in the spring/electromagnet type. The control coil creates a counteracting field to release the brake. Since this design eliminates the inertia of the spring, it will operate at much higher rates than the spring set type, ideal for rapid start/stop applications.

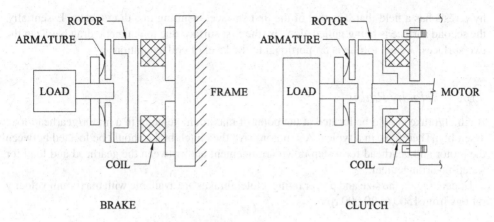

Figure 3.73   Brake and clutch cross-section

One negative aspect of this type of brake is that the power supply for the control coil must be well regulated so that the balancing action between the two fields operates properly and prevents any possible drag.

3.6.1.2   Magnetic Particle

The space between the two surfaces is filled with magnetizable material (typically iron powder). A control coil, when energized, causes the particles to line up to the magnetic field lines created by the coil and lock the two surfaces together. The amount of torque transmitted is directly proportional to the level of magnetization of the particles caused by the current leveling the control coil. This action creates a device in which the transmitted torque is adjustable from zero to the maximum rated value.

3.6.1.3   Eddy Current

An eddy current design consists of two metallic surfaces and a stationary coil. One surface is ferrous and the other is non-ferrous (aluminum, copper or brass). When the coil is energized, its flux links both surfaces. As the ferrous surface rotates in this field it creates internal eddy currents which, in turn, create a secondary flux field generating counter-acting currents in the non-ferrous surface, resulting in a torque. Since relative motion between the two surfaces is necessary to create the torque, this design is not able to be used as a holding brake. However, as in the magnetic particle design, torque is proportional to the level of coil excitation.

3.6.1.4   Hysteresis

The hysteresis design also consists of two surfaces and a coil, except both surfaces are ferrous. When powered, the coil creates a magnetic field in one surface (the input rotor in the clutch version). As it rotates, the field causes magnetic action in the second surface which, due to its hysteresis has a field that lags that of the first surface, resulting in a drag torque. Essentially, the second surface is being pulled along by the first surface due to magnetic drag between the two surfaces. Again, torque is proportional to the level of coil excitation.

3.6.2 Velocity Rating

A clutch/brake should be located at the point of maximum velocity in a motor/gearhead/load assembly. Therefore, in a typical AC motor drive the clutch/brake should be located between the motor and gearhead for a step-down arrangement and between the gearhead and load for a step-up arrangement.

Depending on the size and power rating, clutch/brakes are available with maximum velocity ratings from 1500 to 20 000 rpm.

3.6.3 Torque Rating

Once the velocity rating has been specified, the torque necessary to accelerate or decelerate the load can be determined from:

$$T = (J\theta'' \pm T_L)(SF) = \left(J\frac{\theta'}{t} \pm T_L \right) (SF) \tag{3.118}$$

where $T =$ torque (g cm), $J =$ inertia (g cm s^2), $\theta'' =$ acc/dec (rad s^{-2}), $\theta' =$ velocity (rad s^{-1}), $t =$ time (s), and $T_L =$ load torque (g cm).

Inertia is known or can be calculated or estimated and must include the clutch, brake and gearhead inertia.

The velocity is usually expressed in rpm and converted to rad s^{-1} by multiplying it by $2\pi/60$.

$+T_L$ is used for clutch sizing; $-T_L$ is used for brake sizing.

The safety factor (SF) is an arbitrary value ranging from 1.25 to 2 to account for unknown loads and friction variations. A value of 1.5 is usually a good initial value.

If a gearhead is to be used, Equation 3.118 should be used with respect to all load parameters reflected to the input side of the gearhead and include the efficiency of the gearhead and any mechanical devices on the load side of the gearhead.

The time is dependent on the application and duty cycle. In a clutch application in which the system is started and then allowed to run for many minutes, a relatively large time will be acceptable and result in low torque and the smallest, least expensive clutch, consistent with the clutch supporting any load torque. Also, if a brake is used to provide for a power failure/emergency stop situation, the time should be chosen to stop movement consistent with preventing any harmful result.

Alternately, a clutch/brake assembly used to provide rapid start/stop operation will typically have small times, of the order of tens to hundreds of milliseconds, resulting in a relatively large torque.

The torque calculated by Equation 3.118 is defined as the dynamic torque.

The torque rating listed in clutch/brake catalogues is static torque. Static torque is the torque rating of the device with zero differential velocity between the two surfaces. This would be

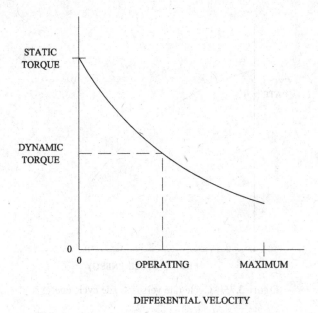

Figure 3.74   Differential torque versus differential velocity

the case, for example, if a clutch is engaged with the motor at zero velocity and then the motor is accelerated.

Catalogs will show a curve or a list for torque versus differential velocity, from zero velocity to the maximum rated velocity. The torque at any differential velocity up to the maximum is the dynamic torque rating at that differential velocity, and is the value that should be used in selecting the clutch or brake (see Figure 3.74).

The ratio of static torque to minimum dynamic torque (at maximum rated velocity) can vary from 1.5:1 to 8:1.

3.6.4   Duty Cycle/Temperature Limits

Clutches and brakes have maximum temperature ratings determined by the insulation characteristics of the control coil and the thermal limitation of the mating surfaces or magnetic material.

In a clutch, there are two main sources of dissipation; the control coil and the clutch slip action as the relative velocity changes from 100 to 0% while transmitting the torque from input to output.

In a brake there is also the control coil dissipation alternating with the braking dissipation while the brake absorbs the kinetic energy from the load in the "power off" type or coincident with the braking energy in the "power on" type.

If both devices are being operated at a low duty cycle, and especially for a brake being operated in an emergency stop and hold application, the dissipation is essentially that of

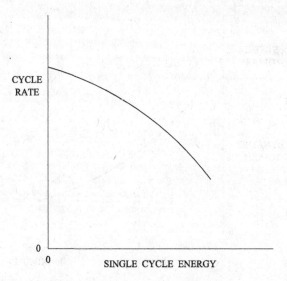

Figure 3.75   Cycle rate versus single cycle energy

the control coil. If operated at the specified voltage, the device will remain well within its temperature rating.

In high cycling rates the load energy being transmitted and absorbed will cause the temperature to rise. Unlike the case for motors, where a published thermal resistance together with a calculated RMS dissipation allows the internal temperature to be determined, the nonlinear nature of a clutch or brake does not permit this.

Instead, manufacturers provide charts showing the maximum permissible cycle rates in two, ways:

(a) Energy per cycle versus cycles per minute (Figure 3.75)
(b) Inertia versus cycles per minute for various speeds (Figure 3.76).

For (a) the energy per cycle, K, must first be calculated by:

$$K = \frac{J\theta'^2}{2} \text{ g cm} \tag{3.119}$$

then the cycle rate determined from the chart.

Note that manufacturers use different units for this expression. Be sure to convert your units to those used on the chart.

For (b), use the total inertia and the maximum operating speed to determine the maximum cycling rate.

Both methods will provide the maximum allowable cycling rate that will result in the devices not exceeding their maximum rated temperature in a 25 °C ambient. Data for elevated ambient temperature is typically not included in the catalogs and should be discussed with the manufacturer.

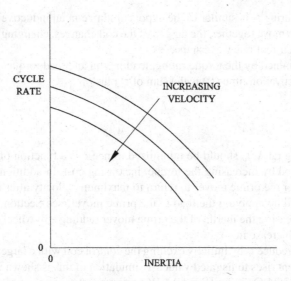

Figure 3.76 Cycle rate versus inertia as function of velocity

Also, although the cycling rate is shown, the duty cycle (the amount of time the device is on and off within one cycle) is not given and should also be discussed with the manufacturer.

Magnetic particle and hysteresis devices are often used in a continuous slip mode, such as maintaining tension in winding applications. In these applications, there will be differential velocity between the input and output at some torque level. Catalog data typically list maximum rated slip dissipation which can be any combination of torque and differential velocity, provided that neither exceeds the maximum rating.

The slip dissipation can be calculated as:

$$P = \frac{TN}{97\,340} \tag{3.120}$$

where P = dissipation (W), T = torque (g cm), and N = differential velocity (rpm).

If the application involves intermittent operation, Equation 3.120 should be used together with the duty cycle to calculate the RMS dissipation.

3.6.5 Timing

The turn on time of a clutch or "power on" type of brake consists of two parts:

(a) the rise time of current in the control coil (t_c) until,
(b) the time for the build up of torque from zero until maximum torque is achieved. (t_t) and the load reaches maximum velocity (clutch) or zero velocity (brake).

The current rise during t_c is similar to the exponential rise in an inductive circuit, except that as the two surfaces move together, the air gap of the coil changes, changing the coil inductance and therefore the current rise waveshape.

Time t_t is determined by the torque rating, load inertia and load torque.

The resultant activation time (t_a) is the sum of t_c plus t_t:

$$t_a = t_c + t_t \tag{3.121}$$

For high cycling rates, t_a should be minimized. Since t_t is a function of the load (fixed), it can only be reduced by increasing the torque, increasing cost. In addition, t_t is a function of the time it takes for the prime mover to return to maximum velocity after the temporary slow down effect caused by applying the load to the prime mover (see Section 5.3 for an example of this effect) Increasing the inertia of the prime mover (adding a flywheel) would also reduce t_t, but again at an increase in cost.

It is possible to reduce t_c by initially exciting the control coil with a larger than rated voltage step until the current rises to its rated value. A simulation of this is shown in Figure 3.77 for a coil rated at 24 V, 76.4 Ω, 730 mH and 7.5 W continuous.

The coil is subjected to 84 V until the current rises to 0.313 A, at which time the voltage drops to the rated value of 24 V, decreasing t_c from 0.05 to 0.003 s.

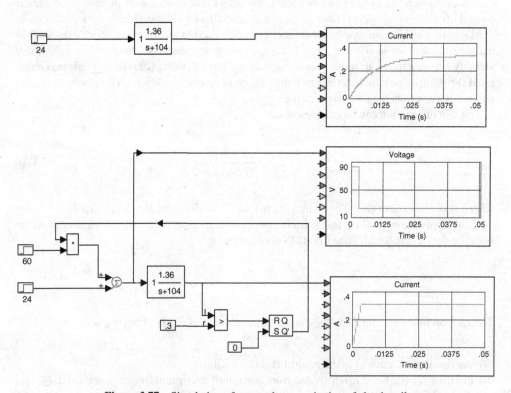

Figure 3.77   Simulation of over voltage excitation of clutch coil

Note that, in this simulation, no attempt has been made to simulate the nonlinear nature of the coil inductance, since in the total time, t_a, it represents a second order effect.

Commercial clutch/brake controllers are available with programmable timers to supply this type of over-voltage action to minimize coil activation time.

Turn off timing is typically less than turn on timing due to the rapid decay of current when the coil is turned off. This is a result of the use of diode, zener diode or MOV spike suppression across the coil switch. If such suppression is not used, current decay will increase and potentially damaging voltage spikes will exist across the coil switch.

The timing of a "brake off/clutch on" or a "clutch off/brake on" sequence should be designed such that there is no or only a short overlap time, determined by the coil on/coil off time, so that the devices do not 'fight' each other.

Simulation of a clutch/brake controlling a conveyor belt is shown in Section 5.3.

3.6.6 Control

Control of a clutch/brake operation can vary from very simple to closed loop as follows.

In all cases a power supply to energize the clutch brake coils and a properly rated prime mover are required. Selection of the prime mover is especially important and must include a detailed analysis of the system dynamics, efficiencies and estimates of load torques and friction effects.

3.6.6.1 Manual – Open Loop

Control is created via manually operated switches, for simple start/stop machine control.

3.6.6.2 Time Based – Open Loop

Positioning control is created using timing circuits, depending on the tolerances of the on/off timing of the clutch and brake and the velocity of the prime mover providing sufficiently accurate positioning.

3.6.6.3 Sensor Based – Closed Loop/Fixed

Sensors placed in the motion path provide position feedback to initiate braking by the controller when the load has reached a preset position. Motion is initiated as a result of a "start" command or the detection of a signal from an associated system.

3.6.6.4 Sensor Based – Closed Loop/Adaptive

An encoder is located to detect load position. The encoder signal is fed back to the controller. The actual position is compared to a nominal commanded value and if the difference is outside an acceptable error band, the timing of the command initiating braking on the next cycle is advanced or retarded, depending on the polarity of the error.

By this method, positioning over a period of cycles is brought within the error band, adjusting for load variations and clutch and brake long term wear.

3.6.7   Brake/System Timing

When using a brake to "hold" the load at zero velocity in a closed loop system, such as in a vertical ball screw system, timing of the turn on and off of the brake is critical.

When turning the brake on, the system should first be brought to a stable zero velocity condition before exciting the brake in order to prevent the system from opposing the brake and possibly stopping short of the commanded position. If t_c is known and repeatable, the brake command could be initiated t_c seconds before the expected time of achieving zero velocity.

When resuming motion, the system should be closed loop enabled to allow it to establish zero velocity conditions before turning the brake off, to avoid a "bump" as motion starts.

3.6.8   Soft Start/Stop

The rapid application of torque to a mechanical assembly by a clutch or brake, essentially a step of torque, can result in noise and vibration possibly leading to premature component failure. Since the transmitted torque in a clutch or electrically on brake is proportional to the control coil current, from zero to maximum rating, applying the coil current as a ramp rather than a step will produce a "soft" action, reducing the vibration.

However, this will increase the start/stop times, resulting in lower maximum cycling rates.

3.7   Servo Couplings

Although a coupling performs a rather simple function, that is, connecting the shafts of two independently bearing suspended assemblies (typically connecting a motor to a load) there are a number of critical items which must be considered in order for the coupling to function reliably without adversely affecting the static and dynamic performance of the complete assembly.

The coupling must not be treated as an afterthought, but it should be considered during the design process along with the mounting tolerances of both bearing structures to provide for best performance.

Couplings have a direct effect on the life of the bearing structures to which they are attached. Using the wrong type or size (rating) coupling will greatly reduce the life of the bearings. During the design phase, estimates should be made of the types of bearings, quality of mounting surfaces and all the classes of misalignment expected in order to select the proper coupling. Coupling selection should not be left as an after thought.

An ideal coupling requires conflicting characteristics:

1. It must be stiff yet have low inertia
2. It must be rigid but be flexible to accommodate angular, axial and parallel misalignment
3. It should have zero backlash and minimal windup.

It should be compliant, yet have a compliance rating that falls above the system bandwidth, especially important in high bandwidth systems.

3.7.1   Inertia

A coupling's inertia must be considered together with the inertial load to which it is connected. When used to connect a motor to a load, coupling inertia can usually be ignored. However, when used to connect a low inertia devise such as an encoder or resolver to a motor, the coupling inertia will usually be of the same order of magnitude as the devise and must be included in any calculations.

3.7.2   Velocity

Depending on the coupling type (see Sections 3.7.6.1 to 3.7.6.6) maximum velocities can range from 4000 to 40 000 rpm. In rapid incremental servo systems, this parameter may be the deciding factor in determining the type of coupling to be used.

3.7.3   Torque

Torque ratings shown in manufacturers' data sheets are usually static ratings. Application notes typically recommend that safety factors of 1.5 to 4 be applied to the maximum torques (usually the peak acceleration torques) that the system will develop. The higher safety factors (3 to 4) are recommended for systems experiencing rapid start/stops or reversals. These safety factors concern both the active portion of the coupling and the method by which it is fastened to the shafting.

3.7.4   Compliance

Assume a simple structure in which two discs are attached to opposite ends of a solid shaft. The basic resonant frequency of this assembly can be calculated as follows:

$$f = \frac{1}{2\pi}\sqrt{\frac{J_1 + J_2}{J_1 J_2 C_t}}\ \text{Hz} \tag{3.122}$$

where: J_1 and J_2 are the inertias of the two discs and

$$C_t = \frac{L}{GI_p}\ \text{rad N}^{-1}\ \text{m}^{-1}, \text{the compliance or torsional flexibility} \tag{3.123}$$

where $L = $ shaft length (m), $G = $ shear modulus of elasticity (N m^{-2}) and $I_P = $ polar moment of inertia $= \dfrac{\pi d^4}{32}$ (m^4).

Note: A term that is the reciprocal of C_t is also used, defined as the torsional stiffness:

$$K_t = \frac{GI_P}{L} \text{ (N m rad}^{-1}) \tag{3.124}$$

There is some confusion among coupler manufacturers concerning the definition of a coupler's ability to resist angular motion when subjected to an applied torque. The terms compliance and stiffness are both used with similar and reciprocal units.

A survey of eight vendors has five of them defining stiffness as torque per angle (N m rad^{-1}) and three as angle per torque (rad N^{-1} m^{-1}). To add to the confusion, they all use the letter "C" as the descriptor for this term.

For this book we will use the following:

$$\text{Torsional compliance} = C_t = \frac{L}{GI_p} \text{ rad N}^{-1} \text{ m}^{-1} \tag{3.125}$$

$$\text{Torsional stiffness} = K_t = \frac{GI_P}{L} \text{ N m rad}^{-1} \tag{3.126}$$

The resonant frequency expression can be written as:

$$f = \frac{1}{2\pi} \sqrt{\frac{J_1 + J_2}{J_1 J_2}} \sqrt{\frac{1}{C_t}} \text{ Hz} \tag{3.127}$$

Thus, by comparing the compliance of various devices, taking into account the square root, their contribution to the resonant frequency can be compared.

It is interesting, theoretically, to compare the compliance of a coupling to the compliance of a solid steel shaft of the same length and diameter.

- A Bellows coupling with: Active length = 8 mm

                                          Bore = 16 mm

                                          $K_t = 1300$ N m rad^{-1}, $C_t = 1/1300 = 7.7 \times 10^{-4}$ rad N^{-1} m^{-1}

- A section of steel shaft: Length = 8 mm

                                          Diameter = 16 mm

                                          $G = 80 \times 10^9$ N m^{-2} for steel

---

$$I_P = \frac{\pi (0.016)^4}{32} = 6.43 \times 10^{-9} \text{ m}^4 \tag{3.128}$$

$$C_t = \frac{0.008}{80 \times 10^9 \times 6.43 \times 10^{-9}} = 1.56 \times 10^{-5} \text{ rad N}^{-1} \text{ m}^{-1} \tag{3.129}$$

$$\text{For the coupling: } \sqrt{\frac{1}{C_t}} = \sqrt{\frac{1}{7.7 \times 10^{-4}}} = 36 \tag{3.130}$$

$$\text{For the steel shaft: } \sqrt{\frac{1}{C_t}} = \sqrt{\frac{1}{1.56 \times 10^{-5}}} = 253 \tag{3.131}$$

The coupling would create a resonant frequency 1/7 of that resulting from the steel shaft for any combination of the two load inertias.

This, of course, is a simple theoretical comparison, but it does highlight the fact that the couplings compliance must be considered when designing an assembly which uses a coupling in order to avoid potential resonances and vibrations.

An assembly consisting of a motor coupled to an encoder via the rear shaft and to an inertial load via the front shaft could easily exhibit two resonant conditions if the couplings are not properly sized for the application.

See Section 4.6.6 for a simulation of a closed loop system showing the effect on performance caused by the compliance of two couplings.

3.7.5 Misalignment

Typically, two assemblies that are coupled will have their locations less than ideal as follows (Figure 3.78):

Parallel:	Also called radial or lateral. The two shafts will be in the same plane but laterally displaced from each other.
Angular:	The two shafts will be in the same plane, but oriented at an angle to each other. Include a safety factor when selecting a coupling to account for change in misalignment as the assembly is operated. Select a 4° rated coupling but design for a maximum of 2° of offset.
Axial:	The two shafts will be in the same plane and on the same axis but will be capable of moving axially with respect to each other. This effect is typically the result of shaft length changes caused by ambient temperature or equipment temperature changes.
Skew:	This term is sometimes used to describe the combined effect of angular and parallel misalignment.

3.7.6 Coupling Types

Ideally, a coupling will be compliant in all three misalignments while having high torsional strength together with no windup or backlash. Actual couplings can by their various designs fulfill only some of these requirements, as described in the following review of the various high performance couplings available. See Figure 3.79.

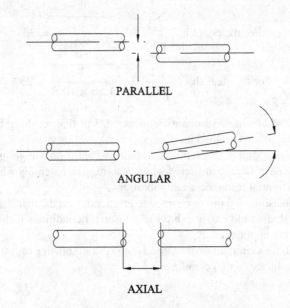

Figure 3.78 Misalignment of coupled shafts

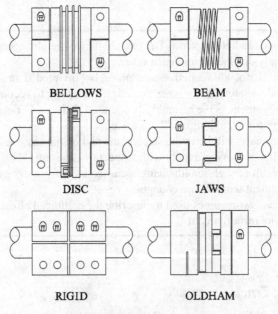

Figure 3.79 Coupling types

3.7.6.1 Rigid

A rigid coupling is simply either a solid tube attached to both shafts with set screws or is a split (two-piece) assembly, the better of the two designs, which can be clamped around both shafts. It offers no compensation for any of the shaft misalignments and, therefore, is rarely used for high speed applications.

A rigid coupling can only be used when precise alignment of both shafts is possible to prevent forces causing bearing, shaft or coupling failure due to thermal expansion and contraction of the coupling caused by heat generated if the assembly operates at high speed or with rapid start/stop modes. However, if misalignment can be tightly controlled, rigid couplings can provide stiff, high torque capacity with zero backlash.

3.7.6.2 Oldham

An Oldham coupling consists of three parts; two hubs and a plastic center disc. The center disc has slots on both sides located 90° apart. These slots mate with tenons located on the hubs in a tight fit, resulting in zero backlash. This construction allows for considerable parallel misalignment, as much as 2.5 mm, but less than 1/2° of angular misalignment and less than 0.127 mm of axial motion. As the coupling rotates, the center disc slides on the tenons of the hubs and, therefore, will wear, resulting in cessation of zero backlash, requiring replacement of the disc. The disc is typically made of various plastic materials, to provide a selection of performance characteristics from tight zero backlash to more flexibility providing vibration and noise absorption. Speed is typically limited to a maximum of 4000 rpm. A positive characteristic of this coupling is that it indirectly provides a machine safety effect. If excessive torque is developed, the disc can fail and motion will stop.

3.7.6.3 Jaws

Also known as Elastomeric or Insert, this coupling also consists of three parts; two hubs and an Elastomeric disc "spider" insert. The spider has multiple teeth that fit between slots alternately on the two hubs. The fit is a press fit resulting in the spider teeth being in compression, providing zero backlash and a certain degree of vibration absorption. This coupling is well balanced, allowing speeds up to 40 000 rpm, but has limited misalignment capability, especially the angular is limited to 1°. Only 0.16 mm of parallel misalignment is provided. The spider can typically be supplied in various materials providing a range of hardness and temperature characteristics.

3.7.6.4 Beam or Helical

This coupling consists of a single cylindrical piece which has been machined with a spiral cut, resulting in a stiff spring-like device. It is available in a single cut (one beam) and multiple cut (two or three beam) design, to accommodate a range of torque and misalignment requirements. Although the single piece construction provides zero backlash, it does have windup and low torsional rigidity. Its "spring" nature provides for angular misalignment and axial motion. In the multiple beam configuration increased torque is provided and two sets of multiple beams allow accommodation of both parallel and angular misalignment. Although typically fabricated in aluminum, they are also available in stainless steel, increasing torque rating and stiffness but also inertia. They typically have a maximum speed rating of 6000 rpm. Some types are available to accommodate up to 90° of angular misalignment. Due to the high windup they are best suited for light load, low torque, non-precise applications.

3.7.6.5 Disc

Disc couplings consist of either two hubs with a thin metallic center disc, or two hubs, a center solid spacer and two thin metallic discs between each hub and the center spacer. The discs are "pinned" rigidly to the hubs, resulting in no backlash and providing up to 5° of angular misalignment and parallel misalignment of up to 0.0254 mm for the single disc and 0.51 mm for the double disc. The two disc design is more accommodating to parallel misalignment since the discs can bend in opposite directions. The stiff discs provide high torsion ratings and can operate up to 10 000 rpm.

3.7.6.6 Bellows

Bellows couplings are made by welding or gluing two hubs to a thin-walled metallic bellows. The bellows is usually made from nickel, using an electrodeposition process, or stainless steel, using a hydroforming process. The electrodeposition process allows the bellows to be made with very thin walls, providing low inertia for high responsiveness. The uniform thin walls of bellows couplings allow them to be compliant in all three misalignments while maintaining torsional rigidity. The nickel bellows together with aluminum hubs provide a low inertia, stiff assembly, ideally suited for wide bandwidth servo applications. They can provide for up to 2° of angular misalignment and up to 0.51 mm of parallel misalignment and axial motion at speeds up to 10 000 rpm.

3.8   Feedback Devices

As described in Section 2.4 "... changes in the feedback term will result in changes in the output directly proportional to the feedback changes. In other words, operation of the system (stability, accuracy, etc.) is mostly dependent on the quality, accuracy, and so on of the feedback element."

Therefore, basic knowledge of the characteristics of various types of feedback devices is necessary in order to select the type best suited to meet the performance specifications of a closed loop motion control system.

The following sections describe the characteristics of various feedback devices, emphasizing optical encoders, resolvers and potentiometers.

3.8.1   Optical Encoders

Optical encoders transform rotary or linear motion into electronic signals for processing by the system controller.

In an optical encoder a light source projects through a focusing graticule onto a masked photo-diode/transistor (see Figure 3.80).

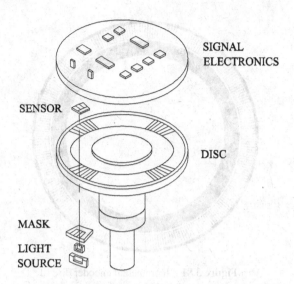

Figure 3.80   Encoder assembly

The light beam is sequentially interrupted by transparent and opaque areas on a rotating disc, resulting in a near sine wave signal. Depending on the design of the disc and associated optics, encoders provide either incremental or absolute shaft angle information. By interpreting the frequency or period and the phasing of these signals, the controller can determine the incremental or absolute position, and the velocity, acceleration and direction of the system motion.

In addition, due to the advent of the brushless motor, encoders now also provide three signals, U, V and W to be used by the motor controller to implement the commutation of the motor stator windings. Used alone, these signals will create a six-step drive. Used together with the A and B signals (see Section 3.8.1.1) and an ASIC chip containing sine tables they will create a sine wave drive.

3.8.1.1   Incremental Optical Encoders

In an incremental encoder the disc has a single track (see Figure 3.81) to effect the light beam interruption and the focusing and masking are designed so that two signals 90 electrical degrees apart are created. These signals are then shaped into square waves by on-board circuits and transmitted to the controller as quadrature signals (see Figure 3.82).

In addition, a third signal, a pulse occurring once per revolution is created to serve as an index mark. These three signals are designated as A, B, and Z.

They are usually created as differential pairs ($A/\bar{A}$, $B/\bar{B}$, $Z/\bar{Z}$) in order to eliminate common noise and random interference problems.

If only the leading, positive going, edge of either the A or B signal is used, the encoder is referred to as operating in the X1 mode; data resolution is equal to the disc resolution. If both positive and negative going edges of either the A or B signal are used data resolution is twice the disc resolution, the 2X mode, and if both edges of A and B are used the data resolution is four times the disc resolution, the 4X mode.

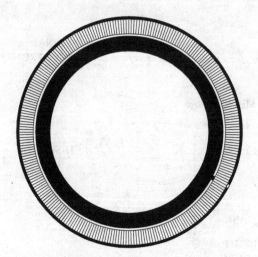

Figure 3.81   Incremental encoder disc

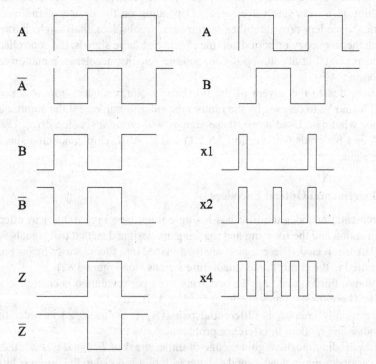

Figure 3.82   A, B, and Z encoder signals

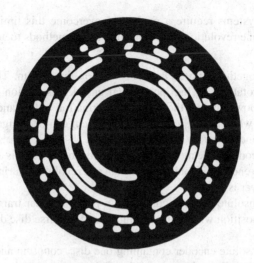

Figure 3.83   Absolute eight bit encoder disc

This is basically a simple implementation of *interpolation*; extracting data at a resolution higher than the basic resolution of the mechanism itself.

Since the raw signals are basically sine waves with a sine/cosine relation, a trigonometric process (similar to that used in resolver R to D conversion) can be used to create even higher resolutions, with multiples as high as twenty times; for example a 1000 count encoder could provide 20 000 counts per revolution.

3.8.1.2   Absolute Optical Encoders

In an absolute encoder there are multiple tracks and companion photo-detectors such that a binary weighted "word" is created for particular positions within one rotation of the disc (see Figure 3.83).

Resolutions as high as 16 bits are available, providing 2^{16} or 65 536 positions per revolution.
The output word is usually available in two standard formats:

- Standard Binary – The normal sequence of binary counting
- Gray Code – A modified counting in which there is only one bit change for each count step.

The Gray Code eliminates the errors that can occur in Standard Binary due to circuit delays, line noise, and so on when some count steps involve the changing of more than one bit.

3.8.1.3   Range/Multi-Turn

Both incremental and absolute encoders have a common characteristic; they only provide positional information for a single rotation of the shaft. Although this is acceptable in velocity systems, positional systems require a method to overcome this limitation for applications requiring more than one revolution of the encoder. Some methods to achieve this are:

- Use the Z signal together with a "home" sensor on the system. The controller can keep track of multiple rotations of the incremental encoder as motion moves away from the home position. If power is lost the position information is lost and the system will have to be re-initialized when power returns. This problem can be mitigated if the controller is equipped with a standby battery power supply.
- Use an absolute encoder while monitoring the number of turns. This approach is also subject to loss of data if the encoder is rotated more than one revolution during loss of power, unless battery backup power is used.
- Use a multi-turn absolute encoder containing a precision gear train and two discs (a fine disc for detecting position within one revolution and a course disc detecting the number of revolutions).
- Use a multi-turn absolute encoder containing one disc, counting and memory circuits and an on-board battery to power the encoder during loss of power.

3.8.1.4　Encoder Specification Summary

The following specifications are a compilation of data from eight major manufacturers:

- Power: 5, 12 and 15 V DC
- Package: Housed and kit
- Diameter: 12 to 152 mm
- Count (incr): 100 to 8192 ppr typical; up to 900 000 ppr special
- Bits (abs): 9, 12, 13, 14, 15, 16, 17 single turn
- Bits (abs): 24 multi-turn (12 bits fine; 12 bits course)
- Code (abs): Natural Binary, Binary Coded Decimal, Gray Code
- Frequency: 300 kHz max.
- Disc: Etched metal, glass/chrome plate
- Velocity: 6000 rpm max typical, 20 000 rpm special
- Commutation: 2, 4, 6, 8 pole 3 phase
- Symmetry: $180 \pm 36°$
- Phase: A/B, $90 \pm 45°$
- Temperature: 100 °C max.
- Interface: See Section 3.8.6.

3.8.1.5　Interface/Data Transmission

Data can be transmitted in either sinusoidal or digital form, in parallel or serial, using a number of circuit topologies:

- Sinusoidal – At any frequency, the sinusoidal signal has the lowest harmonic content and, therefore, requires the lowest transmission bandwidth. Conversion to digital format and interpolation must take place in the controller.

- Digital – Digital formatting takes place in the encoder by squaring the raw sinusoidal signals and transmitting them to the controller in either serial or parallel transmission. Digital is compatible with serial transmission and bi-directional data control, addressing, and so on, especially in multi-encoder bus arrangements. At any data rate, a digital signal requires at least five times the bandwidth as the equivalent sinusoidal signal to maintain signal integrity.
- Parallel – Parallel transmission provides the fastest data transfer rate and is usually the method used for the three channel, A, B, Z incremental encoder, requiring only three twisted, shielded pairs for the data signals. When used for an absolute encoder, it requires a conductor for each bit of the data word, leading to large multiconductor cables and connectors.
- Serial – Serial transmission reduces the number of conductors, but introduces data time delay since each word is transmitted in serial fashion in response to addressing by the controller. Serial is also useful in a bus arrangement for multiple devices, but at the expense of additional delay, depending on the control protocol. Whether this delay affects system performance must be carefully analyzed during initial design to avoid introducing errors in response time, instability or positioning errors.

A number of encoder specific transmission systems are available, including the following:

- Serial – SSI (serial synchronous interface) Data is transmitted to the controller on a bit by bit basis in response to clock pulses supplied by the controller. Data rates available are 100 kHz to 1.8 MHz using differential drivers/receivers for up to 300 m.

  Also available with sine/cosine channels in addition to the digital channel to supply intermediate real time data between the digital absolute words.

  Serial – EnDat (Encoder Data) EnDat is similar to SSI. It has a bidirectional data line that allows the controller to set the zero point, ID number, alarms, and so on. Data rates available are 100 kHz to 4 MHz for up to 150 m. Also available with sine/cosine channels.
- Serial – Hyperface

  Hyperface is only for incremental operation with sine/cosine channels and a bidirectional RS-485 control channel. Operation is asynchronous, no clock.
- Serial – BiSS (Bidirectional Synchronous Serial Interface).

  BiSS has a bidirectional data channel and a clock channel. The controller can interrogate the encoder and then receive the encoder serial word at up to a 10 MHz rate. It does not have sine/cosine capability, eliminating the need for an A to D converter in the controller. It includes a CRC (cyclic redundancy check) character with each data packet to reduce error and provide for limited error correction.

Encoders are also available for operation on standard industrial general purpose buses such as:

- Profibus – European Common Standard EC50170
- DeviceNet – Open DeviceNet Vendor Association (ODCA)
- Interbus.

3.8.1.6 Driver/Receiver Circuits

Driver/receiver circuitry can be broadly divided into the following four categories with respect to the maximum cable lengths that should be observed for reliable operation.

In addition, proper termination, shielding and grounding practices must always be observed:

- TTL – 3 to 5 m
- Open collector – 20 to 30 m
- Differential driver/receiver – 100 m
- RS-422 – 300 m.

3.8.1.7 Signal Frequency/Shaft Velocity/Resolution

These three parameters are related by the following formula:

$$N_{\text{rpm}} = \frac{f\,(\text{counts/s})}{c\,(\text{counts/rev})} \times 60 \tag{3.132}$$

Maximum signal frequency, f, is typically available from 100 to 300 kHz for the basic resolution of an incremental encoder or the LSB for an absolute encoder.

Maximum values for each parameter are usually specified on the data sheet but they are not always mutually compatible.

Example
An encoder has the following specifications:

- Maximum mechanical speed rating, $N = 12\,000$ rpm
- Maximum frequency rating, $f = 180$ kHz
- Line counts, $c = 100$ to 3600 counts per revolution.

Therefore: at a line count of 3600 counts/rev, speed is limited to 3000 rpm at a speed of 12 000 rpm, line count is limited to 900 counts/rev.

3.8.2 Magnetic Encoders

Magnetic encoders are primarily of the linear type. In one form they consist of a plastic strip laminated to a steel backing magnetized in an alternating North/South pattern. A magneto-resistive sensor mounted on the machine carriage detects the field variations that are then converted into incremental output signals.

A second method uses the change in reluctance along the length of either a grooved steel strip or an assembly of precision ball bearings. An inductive detector senses the reluctance change that is then converted to output signals.

Magnetic encoders have coarser resolution than optical types, but are not as fragile and allow larger installation tolerances and withstand shock and vibration disturbances. In addition, they are relatively immune to dirt, dust and especially fluids that can cause errors in optical encoders.

3.8.3 Capacitive Encoders

Capacitive encoders are rugged, low power devices, typically requiring 8 mA at 5V DC compared to the 100 mA required by optical encoders.

They contain an on-board ASIC which generates a 10 MHz oscillator that excites a set of phased transmitters that capacitively couple to a receiver via a shaped rotating disc. The disc modulates the coupling, creating a signal that is proportional to the position of the disc. The signal is decoded into both incremental (A, B, Z) and absolute, up to 12 bits, outputs.

The absolute signal is transmitted via the SPI (serial peripheral interface bus).

Basic resolutions available are 48 to 2048 cycles per revolution (CPR).

Maximum velocity is determined by resolution:

- 30000 rpm up to 512 CPR
- 15000 rpm up to 1024 CPR
- 7500 rpm up to 2048 CPR.

3.8.4 Magnetostrictive/Acoustic Encoders

Magnetostrictive/acoustic encoders use the principle of a sonic wave created by the interaction between a moving magnetic field and a permanent magnetic field located at the position to be measured. The time required for the wave to travel from the magnet location back to the location of the transducer that initiated the moving magnetic field provides a measurement of the magnet location. A novel feature of this system is that more than one position (magnet) can be measured. It measures absolute linear position with an analog output of 0 to 10 V or ±10 V DC, with less than 0.1 mV resolution and linearity of ±0.13 mm up to a 508 mm travel and repeatability of less than 0.08 mm. A to D conversion is performed in the controller.

3.8.5 Resolvers

The resolver is a rotating transformer whose coupling between primary and secondary winding is varied from a maximum to zero by rotating the primary winding mounted on the rotor. It has the appearance of a small motor and has the same rugged construction and high temperature (125 °C) capability.

The resolver is actually one configuration of a family of devices known as synchros that have existed for over 100 years and originally were created for military applications in the area of remote dial indicator/repeater systems, gunfire control, and so on. More recently, they are being used in aircraft and missile control.

The earliest resolvers applied excitation to the rotor winding via a slip ring assembly and typically operated at either 60 or 400 Hz. Due to this mechanical configuration, life and speed were limited. Modern designs include a rotary transformer mounted co-axially on the rotor to supply power to the rotor winding, eliminating the deficiencies of the slip ring version (see Figure 3.84).

The stator windings are physically located 90° apart on the laminated stator assembly (see Figure 3.85).

STATOR

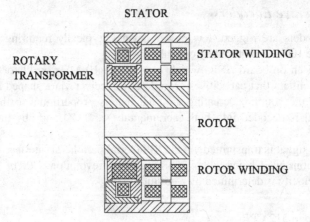

Figure 3.84   Resolver cross-section

3.8.5.1 Resolver Excitation

The rotor has a constant AC reference voltage applied to it with the result that each stator winding has an output voltage that varies from maximum in-phase with the rotor to zero to maximum out-of-phase to zero and back to maximum in-phase.

Since the stator windings are 90° apart, their voltages are proportional to the sine and cosine of the shaft angle as it rotates through 360 mechanical degrees (one revolution). Figure 3.86 shows the waveforms for a resolver with 2 kHz excitation rotating at 3000 rpm.

This action can be expressed mathematically as:

$$E_{rotor} = E \sin \omega t \tag{3.133}$$

$$E_{stator1} = KE \sin \omega t \sin \theta \tag{3.134}$$

$$E_{stator2} = KE \sin \omega t \cos \theta \tag{3.135}$$

where K = coupling constant, typically 0.2 to 1, ω = excitation frequency, typically 2 to 10 kHz, and θ = shaft (rotor) angular position.

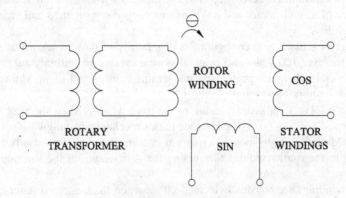

Figure 3.85   Resolver schematic

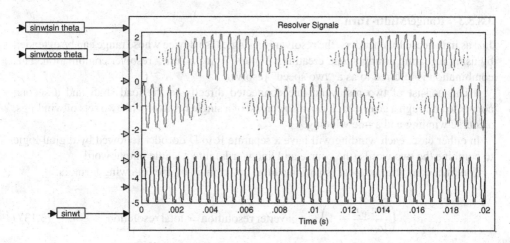

Figure 3.86   Resolver excitation, sine and cosine output

3.8.5.2   Resolver to Digital Conversion

Special resolver to digital (R to D) integrated circuits converts these waveforms into a number of computer compatible signals such as:

- Shaft position; 10, 12, 14 or 16 bit parallel word
- Shaft position; 10, 12, 14 or 16 bit serial word (clocked by external reference)
- Emulated encoder A, B, Z
- Direction (CW, CCW)
- Velocity; 10, 12, 14 or 16 bit signed parallel word
- Velocity; 10, 12, 14 or 16 bit serial word (clocked by external reference)
- Velocity; analog signal.

The majority of converters are "tracking" converters in which trigonometric processing of the two input signals in a Type 2 electronic closed loop servo produces an error signal proportional to the difference between the actual and output angles. This error is continually being driven to zero resulting in the output word being equal to the actual position.

The output angular value will change (increment up or down) an amount as represented by the weight of the LSB.

Example

If the converter is a 10 bit converter, the output will change in increments of:

$$\frac{360 \times 60}{2^{10}} = 21 \text{ arc min} \tag{3.136}$$

One important characteristic of the tracking conversion method is that it is basically implemented by processing the ratio of the two inputs and, therefore, any noise or waveshape distortion occurring on both signals will be canceled by the ratio metric process.

3.8.5.3   Range/Multi-Turn

Just as in the case for encoders, the resolver is a single turn device whose range can be extended by using a gearing arrangement, creating a "course" plus a "fine" resolver combination. This combination is referred to as a "two speed" resolver.

It can consist of two resolvers; one connected directly to the load shaft and a second connected through a gearhead. It can also consist of a single resolver with two sets of windings, a course winding and a fine winding.

In either case, each winding will have a separate R to D decoder followed by digital logic to combine the two words (the course and fine words) into a single output word.

The final combined output resolution can be determined by the following formula:

$$\left(\frac{\log_{10} n}{\log_{10} 2} - 1\right) + \text{converter resolution} = \text{total resolution} \tag{3.137}$$

For example: Assume a 16:1 ratio and a 14 bit converter

$$\text{Total resolution} = \left(\frac{\log_{10} 16}{\log_{10} 2} - 1\right) + 14 = \left(\frac{1.204}{0.301} - 1\right) + 14$$

$$= (4 - 1) + 14 = 17 \text{ bits} \tag{3.138}$$

Then the minimum angular change in the output will be:

$$\frac{360 \times 60 \times 60}{2^{17}} = 10 \, \text{arc s} \tag{3.139}$$

3.8.5.4   Accuracy versus Resolution

As shown above, system resolution is just a function of the resolution of the converter and the course/fine ratio in a two speed system.

Of major importance is the system accuracy that is made up of:

- The accuracy of the resolver, in a single speed system
- The accuracy of the fine resolver in a two speed system (R_f)
- The accuracy of the fine converter (RD)
- The ratio in a two speed system (n)
- The backlash in a mechanical two speed system (B)
- The gearing cyclic errors in a two speed system (G).

$$\text{Total Accuracy} = B + G + \frac{R_f + RD}{n} \tag{3.140}$$

3.8.5.5 Resolver Specification Summary

• Power:	2 to 26 V RMS, 400 Hz to 10 kHZ:
• Package:	Housed and Kit
• Diameter:	30 to 150 mm
• Shaft Size:	3 to 10 mm
• Bore Size:	6 to 100 mm
• Trans. Ratio:	0.25 to 1.75
• Speed:	1X, 2X, 3X, 4X, 5X; Multi-Speed up to 36X
• Error:	±3 arc min to ±30 arc min
• Temperature:	200 °C max.

3.8.5.6 Converter Specification Summary

• Power:	5 V DC ±5% @ 20 mA nominal; and V DC max
• Bandwidth:	1 to 2.4 kHz
• Angular Accuracy:	±22 arc min max
• Resolution:	10, 12, 14, 16 bits
• Tracking Rate:	750 rev/s max @ 6.1 MHz clock
	1000 rev/s max @ 8.2 MHz clock
	1250 rev/s max @ 10.2 MHz clock
• Acc. Error:	30 arc min @ 10 000 rev/s^2 @ 8.2 MHz clock
• Settling Time:	5.2 ms to within ±11 arc min for a 180° step input change
• Vel. Output Accuracy:	2 LSB
• Vel. Output Linearity:	1 LSB
• Enc. Output Count:	256 for 10 bit resolution
	1024 for 12 bit resolution
	4096 for 14 bit resolution
	16384 for 16 bit resolution

3.8.6 Inductosyn

The Inductosyn, like the optical encoder, is available for both rotary and linear applications. Functionally it operates similar to a resolver in that it has a rotor, powered by a high frequency sine signal and two stators that are separated by 90°.

Physically, both the rotor and stator have a continuous printed circuit pattern in a square wave shape in which the square wave repeats many times in 360 mechanical degrees.

Each square wave cycle operates like a resolver, but the square wave patterns are connected in series, averaging out any mechanical deviations from the ideal angular or linear pitch resulting in extremely low accuracies (see Section 3.8.5.6).

The Inductosyn can be thought of as a very high pole count (speed) resolver with as many as 2000 poles/1000 speed.

Connection to the rotor is either via a flexible twisted pair, for limited rotation, or via a rotary transformer, similar to that in a resolver. Inductosyns are available in either incremental or absolute configurations.

An interesting application of the Inductosyn is its use in very low speed velocity systems.

Example

In an application for sidereal experiments, a 512 speed unit was used together with a 16 bit decoder, a brushless motor, linear sine wave amplifier and a 31 457 Hz crystal oscillator (divided by 81) to create a phase-lock velocity system operating at one revolution per day with 0.0009% accuracy.

3.8.6.1   Inductosyn Specification Summary

- Power: 2 V @ 2.5 to 100 kHz:
- Package: Housed and Kit
- Diameter: 7.62 to 40.6 cm (rotary)
- Length: 25 to 36 m (linear)
- Speed: 50 to 1000 (rotary)
- Cycle Length: 2 mm (linear)
- Accuracy: ±1 arc s to ±15 arc s (rotary)
- Accuracy: ±0.0025 mm (linear)
- Trans. Ratio: 0.0002 to 0.05.

3.8.7   Potentiometer

The potentiometer is one of the oldest feedback devices used in motion control systems. It is available in both rotary and linear versions and is typically referred to as a "precision poten-tiometer" to distinguish it from the control type of potentiometer used on various manually operated control panels.

The resistive element in a potentiometer is fabricated from high resistance wire, a wire wound potentiometer, or from various plastic or ceramic compounds.

It is a three terminal device, with a voltage source connected to the extreme ends of the resistance element and a third terminal connected to a wiper arm which slides along the resistive element, providing an output voltage proportional to the distance the wiper moves from the end of the element.

They are available in both single and multi-turn configurations. They are relatively inex-pensive and are true, stand alone absolute positional measuring devices, requiring no auxiliary electronics other than the A to D converter in any associated controller. They are unique in that they require no battery back up to provide total power failure recovery, even if moved while power is off.

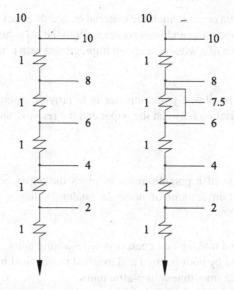

Figure 3.87 Wire wound potentiometer schematic

3.8.7.1 Wire-Wound Potentiometer

The wire-wound potentiometer consists of a coil of wire wound around a mandrel with a wiper making contact with successive locations on each coil as it moves along the winding. It can be thought of as a series of resistors connected in series with the wiper making contact with each series connection.

This construction results in the output as a stair step of voltage as the wiper moves from coil to coil. The output is not truly linear but consists of a large number of discrete steps.

The potentiometer then has a discrete resolution determined by the number of coils.

For example: If there are 100 turns, then the output will have a resolution of 1/100 or 1% of the maximum value. This is analogous to the number of slots in the disc of an optical encoder.

The resolution can be decreased by using a larger number of turns of finer wire but this requires a finer wiper design and can result in accelerated wiper/wire wear, decreasing component life.

One problem is that as the wiper moves from coil to coil, it can bridge two contact points, momentarily shorting out a coil and causing a momentary intermediate output value as illustrated in Figure 3.87.

3.8.7.2 Plastic Potentiometers

Plastic potentiometer is a general term covering three main materials used to form the resistive unit:

- Conductive plastic – Various compounds used to coat a backing material that are processed and shaped to provide the resistivity required, smoothest performance and lowest electrical noise

- Cermet – A combination ceramic/metallic material coated on a backing plate Provides stable performance over temperature and high resistance into the megohm range
- Hybrid - A combination of a wire-wound coil impregnated with a high temperature conductive plastic.

The main objective of the plastic potentiometer is to provide essentially infinite resolution together with very low friction between the wiper and the resistive surface.

3.8.7.3   Noise

Since the basic operation of a potentiometer involves the wiper being in contact with the resistive element, a certain amount of noise or random voltages will be generated from conditions such as:

- The wiper breaking and making coil contact in wire-wound units
- Wiper vibration caused by both internal and external mechanical influences
- Variation in the surface smoothness of plastic units
- Surface contamination in plastic units
- Variation in wiper contact resistance due to wiper pressure variation.

In addition, being electrically unbalanced, potentiometer wiring can be susceptible to external noise sources. This can be minimized by using as low a resistance unit as possible, consistent with the dissipation rating of the potentiometer. In addition, placing a load on the output can lower impedance but this will cause calibration errors, as discussed in Section 3.8.7.6.

3.8.7.4   Tolerance

The body resistance of potentiometers will have tolerance, as high as 10%. Fortunately, except for determining the excitation current and dissipation of the unit, the tolerance does not affect performance since the application is a voltage divider in which the output voltage is only a percentage (from 0 to 100%) of the total motion.

3.8.7.5   Linearity

Linearity is also known as conformity, that is, the deviation of particular potentiometers transfer function from the theoretical transfer function.

For a regular rotary potentiometer the transfer function is:

$$E_{\text{OUT}} = \theta E \tag{3.141}$$

where E_{OUT} = the voltage at the wiper terminal, E = the voltage powering the potentiometer,. and θ = the fractional position of the wiper, varying from 0 to 1.

The plot of this transfer function is a straight line from:

$$E_{\text{OUT}} = 0 \ @ \ \theta = 0 \text{ to } E_{\text{OUT}} = 1 \ @ \ \theta = 1$$

- Absolute Linearity: If actual E_{OUT} versus θ is plotted against this straight line, the maximum deviation from the straight line, expressed as a percentage of the maximum, E, is the absolute linearity.
- Independent Linearity: If the best straight line is plotted through the actual data, then the maximum deviation from this straight line, expressed as a percentage of the maximum, E, is the independent linearity.

    The end points of the resulting straight line will not typically go to $E_{OUT} = 0$ @ $\theta = 0$ and $E_{OUT} = 1$ @ $\theta = 1$.
- Zero Based Linearity: Similar to independent linearity except the straight line must start at $E_{OUT} = 0$ @ $\theta = 0$.

Note that the independent linearity method will typically create the smallest deviation and is the one usually listed on product data sheets.

3.8.7.6   Loading

The potentiometer will typically have a load resistance connected between the wiper output terminal and ground, as shown in Figure 3.88.

The load resistance, R_L is in parallel with the lower portion, θR, of the potentiometer and in series with the upper portion, $(1 - \theta) R$. The output voltage, E_{OUT} will then be:

$$E_{OUT} = \frac{E}{(1 - \theta) R + \dfrac{\theta R R_L}{\theta R + R_L}} \times \frac{\theta R R_L}{\theta R + R_L} \tag{3.142}$$

$$\text{Define: } n = \frac{R}{R_L} \tag{3.143}$$

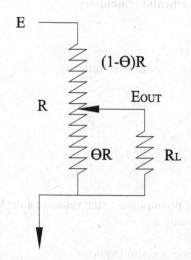

Figure 3.88   Potentiometer loading diagram

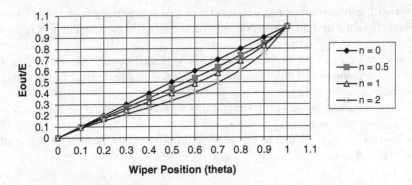

Figure 3.89   Potentiometer output ratio/wiper position versus loading

then

$$\frac{E_{OUT}}{E} = \frac{\theta}{1 + \theta n(1 - \theta)} \tag{3.144}$$

Figure 3.89 shows a plot of this expression for four values of n.

When $n = 0$, $R_L = \infty$: the expression is the straight line. As R_L decreases, the error between the actual E_{OUT} and the ideal, unloaded value increases.

In general, it is best to use the lowest possible resistance value for R and a value for R_L at least ten times R ($n \leq 0.1$)

Also, Equation 3.144 could be implemented in a controller software routine to calculate the true value of θ as a function of E_{OUT}, E and n.

3.8.7.7   Potentiometer Specification Summary

- Power: 1 to 10 W
- Resistance: 50 Ω to 500 kΩ
- Type; Wire-wound, Cermet, Conductive plastic
- Linearity: ±0.15 to ±10%
- Size (dia.): 14 to 50 mm
- No. of Turns: 1, 3, 5, 10
- Turns Life: 100 K to 10 M.

3.8.8   Tachometers

Tachometers provide a signal proportional to the velocity of the load.

They have two main applications:

- To provide feedback in velocity control systems
- To aid in stabilizing positioning systems.

3.8.8.1 DC Brush Tachometers

These tachometers resemble small DC motors, with a permanent magnet stator and a wound rotor. When rotated, the coils on the rotor have alternating voltage induced in them which is converted into DC by the action of a set of commutator bars and brushes.

The result is an output voltage with a predominant DC component proportional to the velocity plus a ripple component whose frequency is proportional to the velocity and the number of commutator bars.

The DC calibration of tachometers can range from 1 V/1000 rpm to 50 V/1000 rpm, with special designs having a calibration as high as 200 V/1000 rpm for low speed applications.

The ripple component can vary from 1.5 to 3% of the DC component at any speed, but is typically specified above some minimum speed such as 50 or 100 rpm.

The ripple frequency, depending on the number of commutator bars, can range from 10 to over 200 cycles/rev.

This ripple frequency can cause the controlled velocity to have ripple content if the frequency is inside the bandwidth of the system.

For example: Figures 4.14 and 4.15 show a velocity control system using a tachometer with 7 V/1000 rpm, 12 commutator segments and 3% ripple.

In Figure 4.14 the system is operating at 1000 rpm (105 rad s^{-1}) and the output velocity shows no sign of the tachometer ripple

In Figure 4.15 the system is operating at 50 rpm (5.25 rad s^{-1}) with a decided ripple content in the output velocity.

It is both the ripple frequency and the low value of the DC component that can cause design problems with tachometer-based systems requiring operation over a wide speed range:

For example: A 7 V/1000 rpm tachometer will have 14 V output at 2000 rpm but only 0.14 V at 20 rpm. Theoretically, with sufficiently high enough loop gain operation should be satisfactory. But, in addition there will be brush noise, ground drops, and so on that are of the same order of magnitude as the DC component.

Solutions to low speed brush tachometer based velocity system problems are not straightforward.

In this example, changing to a 45 V/1000 rpm tachometer would raise the low speed output to 0.9 V, but then the 2000 rpm output would become 90 V, well above the typical ±15 V circuit supply voltages.

The ripple could be low-pass filtered, but would require a break point well within the system bandwidth and would lead to system instability.

Changing to a tachometer with a larger number of commutation bars could solve the problem, but will usually require a larger and more expensive device.

3.8.8.2 DC Brushless Tachometers

Just as the case for the brush tachometer, a brushless tachometer resembles a brushless motor with the shaft driven by the load. Since the rotor has the permanent magnets, the output derives from the stator windings.

Again the output amplitude is proportional to the velocity and will be either sinusoidal or trapezoidal depending on the winding and magnetic circuit design.

Since the output comes from the stator there are no brushes and, therefore, no brush noise, but the low output at low speed problem still exists.

In addition, external circuitry is required to "rectify" the three phase output to obtain a DC signal. This signal will contain ripple whose frequency is proportional to six times the pole pair count. The polarity of the signal must be determined by sensing the phase relationship of the three phases which, during a velocity reversal, will involve the time required for the system to decelerate to zero velocity and reverse. At low velocities, incorrect polarity can exist while the phase detection circuit "catches up" to the actual polarity, resulting in a momentary positive feedback condition.

Alternately, the signal frequency or period can be measured, similar to the method used with encoder feedback. However, this method is also subject to the low amplitude problem at low velocities, whereas in an encoder implementation the amplitude of the raw detected sinusoidal signal is proportional to position and is independent of velocity.

References

[1] Menta, S. and Chiasson, S.J. (1998) Nonlinear control of a series DC motor, theory and experiment, *IEEE Transactions on Industrial Electronics*, 45 (1), 134–141.

[2] Hanselman, D. (2003) Brushless Permanent Magnet Motor Design, The Writers Collective, Cranston, Rhode Island.

[3] Yedamale, P. (2003) Brushless DC (BLDC) Motor Fundamentals, AN885, Microchip Technology, Inc.

[4] Fleming, D. (October 2002) Inside angle on sine-driven BLDC motors, Motion System Design, pp. 27–30.

[5] McCormick, M. (June 1990) Sinusoidal Commutation Cuts Torque Ripple in Brushless DC Motors, PCIM, pp. 22–25.

[6] Infranor Inc. (April 1997) Beating torque ripple in brushless servos, Machine Design, pp. 118–120.

[7] Wilson, M. (November/December 1994) Unraveling the mysteries of linear motors, Motion, pp. 26–31.

[8] Wilson, M. (May/June 1995) What is a linear motor, Motion, pp. 21–25.

[9] Stämpfli, H. (January 2004) Brushless linear motors, Part 1, Motion System Design, pp. 28–32.

[10] Stämpfli, H. (April 2004) Brushless linear motors, Part 2, Motion System Design, pp. 28–32.

[11] Novotnak, R., Sobek, R. and Botos, S. (February 1997) Linear motors: How to beat the heat, Power Transmission Design, pp. 51–54.

[12] Condit, R. and Jones, D. (2004) Stepping Motor Fundamentals, AN907, Microchip Technology Inc.

[13] Yorkel, R. (1973) Open Loop Control of Step Motors. Incremental Motion Control Systems and Devices Symposium, pp. 149–230.

[14] Hurst, S. (May 1993) Chopper Amplifier IC Increases High-Speed Torque of Bipolar stepper Motor, PCIM, pp. 19–26.

[15] Kuo, B. and Singh, G. (1973) Damping Methods for Step Motors. Incremental Motion Control Systems and Devices Symposium, pp. H1–H13.

[16] Nordquist, J. (December 1996) Controlling Resonance and Damping in Hybrid Step Motors, PCIM, pp. 36–51.

[17] Leenhouts, A. (February 1992) Fundamentals of Step Motor Control-Part 3 Step Timing for High Performance Open Loop Control, PCIM, pp. 21–26.

[18] Goodrick, S. (April 1997) Electronic Damping Cures Step Motor Resonance-Part 1. Step Motor Characteristics, PCIM, pp. 20–28.

[19] Goodrick, S. (May 1997) Electronic Damping Cures Step Motor Resonance-Part 2. Damping Techniques, PCIM, pp. 32–43.

[20] Endo, K. and Kobayashi, F. (July 2001) Quick stop, Motion System Design, pp. 21–24.

[21] Leenhouts, A. (May 1992) Fundamentals of Step Motor Control-Part 4 Control of Stepping Motors-Microstepping and Torque Feed forward, PCIM, pp. 32–36.

[22] Beauchemin, G. (October 2003) Microstepping myths, Machine Design, p. 86.

[23] Ohm, D. (March/April 1991) Field oriented control of induction motors, Motion, pp. 3–14.

[24] Wilson, C. and Hansen, R. (April 1991) Practical techniques for implementing vector control, Motion Control, pp. 22–25.

[25] Neacsu, D. Space Vector Modulation-An Introduction, IECON2001, pp. 1583–1592.

[26] Parekh, R. (2003) AC Induction Motor Fundamentals, AN887, Microchip Technology Inc.

[27] Bowling, S. (July 2006) DC motor impersonators, Machine Design, pp. 74–81.

[28] Ross, D., Theys, J. and Bowling, S. (2007) Using the dsPIC30F for Vector Control of an ACIM, AN908, Microchip Technology Inc.

[29] (January 2002) ADSP-21990: Reference Frame Conversions, AN21990-11, Analog Devices.

[30] Doebelin, E. (1998) *System Dynamics*, Marcel Dekker, New York, NY, pp. 295–300.

Additional Readings

Brown, G. and Ellis, G. (December 1998) How low-cogging servomotors deliver better machine response, Machine Design, pp. 106–112.

Cheng, M. and Rejass, R. (October 2008) How to take vibration out of step motors, Machine Design, pp. 80–82.

Cooksey, L. (July/August 1999) Driving Process Control With Vector Technology, Motion Control, pp. 24–27.

Fleming, D. (February 2000) Synchronizing Hall-effect devices for brushless motors, Machine Design, pp. 76–78.

Floresta, J. (October 1998) Driving linear motors, Power Transmission Design, pp. 31–34.

George, J. (1999) Comparing the attributes of linear drives, Machine Design, pp. 122–124.

Hendershot, J. (April 1991) A comparison of AC, brushless and switched reluctance motors, Motion Control, pp. 16–20.

Heilig, J. (April 2001) How linear motors measure up, Motion System Design, pp. 25–27.

Ingalls, K. (April 1992) AC vector drives offer alternative to DC motor systems, Motion Control, pp. 25–27.

Jones, D. (November/December 1996) Which motor/drive? Motion Control, pp. 20–24.

Kiner, P. (August 2001) Straighten up and go linear, Machine Design, pp. 96–98.

Leenhouts, A. (September 1991) Fundamentals of Step Motor Control-Part 1 Motor and Driver Selection, PCIM, pp. 42–44.

Leenhouts, A. (November 1991) Fundamentals of Step Motor Control-Part 2 System Response to Single Step Sequences, PCIM, pp. 28–32.

Manea, S. (2009) Stepper Motor Control with dsPIC^{R}DSC, AN1307, Microchip Technology Inc.

Mazurkiewicz, J. and Ohm, D. (November 1991) AC motors for servo positioning Part 1, Motion Control, pp. 36–39.

Mazurkiewicz, J. and Ohm, D. (January 1992) AC motors for servo positioning Part 2, Motion Control, pp. 15–20.

Ring, J. (June 2000) Motor magnetics, Motion Control, pp. 45–46.

Scimia, A. (September/October 1995) Sensorless vector control, Motion Control, pp. 16–18.

Sitapati, K. and St. Germain, R. (June 2000) Reducing cogging torque in brushless motors, Machine Design, pp. 96–104.

Stepper Catalogue, Lin Engineering, Santa Clara, CA 95054.

Tal, J. (Spring 1997) Control modes of step motors, Motion, pp. 12–17.

Thzynadlowski, A. (2001) Control of Induction Motors, Academic Press, San Diego, CA, pp. 92101–4495.

Trilogy Linear Motor Engineering Reference Guide, Parker Hannifin Corp., pp. 1–10.

Zajac, P. (December 2010) Linear motor benefits in systems design, Design News, pp. M10–M13.

4

System Design

4.1 Position, Velocity, Acceleration, Jerk, Resolution, Accuracy, Repeatability

The four basic parameters describing the activity in a motion control system are:

Position

Velocity

Acceleration

Jerk

Once specified, as system design progresses these four parameters have to be reviewed and possibly modified in the light of the specifications and cost of the various components chosen.

4.1.1 Position

A motion system translates or rotates a load from a starting position to a commanded target position with a required precision. The precision of the move is the combined result of three characteristics:

Resolution

Accuracy

Repeatability

Position has the units of: cm (in) for translation, symbol S
rad for rotation, symbol θ

Electromechanical Motion Systems: Design and Simulation, First Edition. Frederick G. Moritz.
© 2014 John Wiley & Sons, Ltd. Published 2014 by John Wiley & Sons, Ltd.
Companion Website: www.wiley.com/go/moritz

4.1.1.1 Resolution

Resolution is the smallest positional increment that the system can achieve. It must take into account such items as encoder resolution, system friction, backlash, hysteresis, lead screw pitch, and so on.

Care must be taken to evaluate the resolution of each component contributing to the system and its effect on the overall system design.

For example, in an encoder-based system, higher resolution can be achieved with higher encoder line count. However, this will reduce the maximum system velocity due to the maximum frequency response limit of the encoder (Section 3.8).

In a resolver-based system, the resolution of the R/D converter (8, 10, 12, 16 bits) which will also determine the maximum velocity, must be considered.

Theoretically, a potentiometer creates an analog signal with infinite resolution. However, the potentiometer signal will pass into the control computer via an A/D port also with 8, 10, 12 or 16 bit resolution plus the potentiometer itself will have a linearity error.

In a stepper motor system with micro-stepping operation, a command to make one micro-step may result in no motion if the system friction is larger than the commanded incremental torque (Section 3.1).

4.1.1.2 Accuracy

Accuracy specifies the maximum difference between a commanded position and the actual position at the end of the move.

Among the items determining system accuracy are encoder resolution, stepper resolution, lead screw lead error, periodic component error, ball nut backlash, thermal variations, and so on.

Example

In a lead screw design, it may initially be decided to use the encoder on the motor shaft together with the pitch of the screw to achieve the required resolution and accuracy.

If this will not provide the desired result, then the use of a linear encoder to measure actual load position may be required. However, this places the lead screw backlash inside the servo loop and may cause stability problems requiring a more sophisticated compensation scheme.

In order to understand the difference between resolution and accuracy, consider a typical 1000 line encoder. The signal from the encoder divides one revolution into 1000 parts; it "resolves" one revolution into 1000 parts, that is, it has a *resolution* of 1000.

The ideal angular distance between parts is $360 \times 60/1000 = 21.6$ arc min. If the signal from the encoder (a square wave) is at a positive going edge, then the distance to the next or previous positive going edge is 21.6 arc min. Therefore, the ideal accuracy of the 1000 line encoder is 21.6 arc min. However, there can be a manufacturing tolerance of as much as ± 0.5 arc min in the symmetry of the pattern on the encoder disc, resulting in an *accuracy* of 21.6 ± 0.5 arc min. In addition, there is an error in the 50/50 duty cycle (the positive going edge to the negative going edge) of the signal plus the error in the quadrature timing between the A and B signals.

A stepper motor operating in full step mode has a resolution of 200; that is, it divides (resolves) a single revolution into 200 parts, or 1.8° (108 arc min) per step. Stepper motors can be obtained with an accuracy of 1 to 5% per step non-cumulative, depending on their quality and cost, with 3% being typical. Therefore, for a 3% device, each step can be 108 ± 3.24 arc min.

Lead screw resolutions are typically given in distances traveled per revolution; mm/rev or in/rev. However, tolerances are not as easily obtained. They are typically given as mm/rev, in/rev, in/ft, mm/300 mm. Ideally, the tolerance should be specified in mm/degree for each lead, which would allow a motor (DC or stepper) /lead screw combination to be analyzed for total accuracy. Using data that are available, a lead screw with a lead of 2 mm/rev and a tolerance of $\pm 0.000\,254$ mm/rev being driven by a full step stepper motor will experience a translation of 0.01 mm $\pm 0.0635\%$ per step. To this must be added the step accuracy of the motor to determine the total motion accuracy.

Because of the less than complete and/or accurate information available for the various components, it is best to avoid designing a system that just relies on the minimal resolution defining the system. A good rule of thumb is to select components that can supply a resolution of three to five times smaller than the required positional error.

4.1.1.3 Repeatability

Repeatability is a measure of how well a system returns to a specific position over a number of identical moves. It is a statistical quality gathered over those moves. It can be specified as unidirectional or bidirectional. Unidirectional is easier to achieve since backlash does not affect the operation, but motion to the commanded position must always occur from the same direction, increasing total motion time. Bidirectional requires close control of backlash, pre-loads, belt/pulley/coupling compliance, and so on.

Figure 4.1 shows a graphical example of accuracy and repeatability.

4.1.2 Velocity

Velocity expresses the speed of the load, or the rate at which the load is changing its position.

Velocity has the units of: cm s^{-1} (in s^{-1}) for translation,   symbol $S' = dS/dt$

rad s^{-1} (rpm) for rotation,   symbol $\theta' = d\theta/dt$

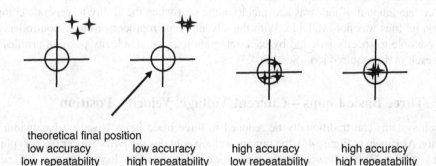

theoretical final position

| low accuracy | low accuracy | high accuracy | high accuracy |
| low repeatability | high repeatability | low repeatability | high repeatability |

Figure 4.1 Graphical example of accuracy and repeatability

Specifying the maximum system velocity and in turn the maximum motor velocity, together with the anticipated supply voltage will contribute directly to determining the voltage constant of the motor (K_e). This then results in the motor torque constant (K_t), which together with the acceleration requirement will determine the peak motor current.

The maximum velocity will also place an upper limit on the allowable encoder count or resolver resolution. Operation over a wide velocity range of a low inertia system can lead to problems at low velocity where the encoder frequency can fall within the bandwidth of the system and cause perturbations of the load velocity at the encoder or resolver frequency.

4.1.3   Acceleration

Acceleration expresses the rate at which the velocity is changing.

Acceleration has the units of: cm s^{-2} (in s^{-2}) for translation,   symbol $S'' = d^2 S/dt^2$
rad s^{-2} for rotation,   symbol $\theta'' = d^2\theta/dt^2$

Acceleration will determine the torque/force required from the motor and if K_t has been determined as a result of velocity considerations it will determine the peak current required from the motor amplifier. Alternately, if a certain acceleration is required and a certain peak current is available, they will determine K_t, which in turn will determine K_e which will then determine the maximum possible velocity.

4.1.4   Jerk

Jerk expresses the rate at which the acceleration is changing

Jerk has the units of: cm s^{-3} (in s^{-3}) for translation,   symbol $S''' = d^3 S/dt^3$
rad sec^{-3} for rotation,   symbol $\theta''' = d^3\theta/dt^3$

Jerk is not mentioned in the literature until the late 1900s. It essentially describes the smoothness or lack thereof (jerkiness) when there is a change in acceleration. Low jerk has become important in semiconductor manufacturing, for example, where low inertia loads must be started and stopped with minimal vibration.

A certain amount of jerk was acceptable in the past when the ability to generate complex motion profiles was not available. With the advent of microprocessor-based controllers it is now possible to specify jerk and by back integration arrive at velocity/position profiles that will result in the required jerk (Section 4.3).

4.2   Three Basic Loops – Current/Voltage, Velocity, Position

Motion systems can traditionally be reduced to three basic loops based on bandwidth and activity. As shown in Figure 4.2, the innermost loop contains the prime mover (motor) plus its drive amplifier. Next is the velocity loop in which a measure of the load velocity is fed back to the velocity command to create the velocity error. The final loop is the position loop in which the position command is compared to the load position to create the main error signal.

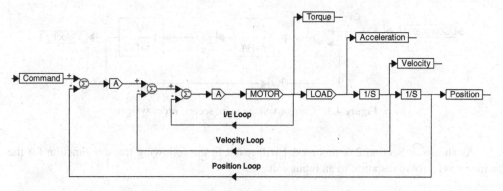

Figure 4.2   Block diagram of three basic loops

Traditional implementation of the loops has been done with separate feedback devices; tachometers or resolvers for velocity sensing and potentiometers or resolvers for position sensing.

The advent of computer (microprocessor) control has virtually eliminated these separate loops and replaced them with the use of a single feedback device, typically an encoder, from which load position and velocity can be determined. Various algorithms are then used to stabilize the system by implementing various forms of compensation (lead/lag, PI, PID, etc.). However, reviewing the "three loop" approach helps the designer develop a "feel" for the functionality of the system, especially in the area of the effect that various component parameters have on system performance.

In addition to these three loops, supplementary feed-forward paths (velocity or acceleration dependent) are sometimes used to improve load dynamic response and reduce positioning time (see Section 4.4).

4.2.1   Current/Voltage Loop

In most texts, the innermost loop is typically referred to as the current loop, since it is the current which produces torque, which in turn produces acceleration resulting in load motion. However, this assumes a pure current amplifier, that is, an amplifier in which an input voltage command produces a proportional output current to the motor. In addition the load is assumed to be purely inertial.

With modern simulation techniques, it is possible to simulate the motor electrical characteristics together with the current feedback path, resulting in a more accurate analysis of the amplifier/motor combination.

In addition, as discussed in the following sections, it is sometimes more efficient to drive the motor with a voltage amplifier in which the error voltage is simply amplified to the appropriate amplitude and power level and the current is determined by the BEMF and the electrical constants (resistance and inductance), as shown in Figure 4.3 for a second order system.

It would be more accurate and appropriate to call this innermost loop the "motor/amp loop" or the "current/voltage loop".

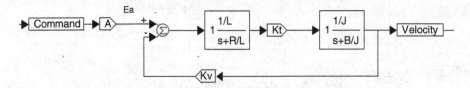

Figure 4.3 Current/voltage loop; second order system

As shown in Section 2.6, this model will result in the following transfer function for the motor velocity in response to an input voltage:

$$\frac{\theta'}{E_A} = \frac{K_T/JL}{s^2 + \dfrac{(JR+BL)}{JL}s + \dfrac{(K_T K_V + BR)}{JL}} \tag{4.1}$$

Note that this complete model of the motor is itself a single closed loop system by virtue of the BEMF feedback, and as such can be evaluated using the standard techniques. Note also that in this voltage driven model, the amplifier is outside the loop and at this level does not contribute to the motor's dynamic response.

Using typical motor data sheet information, this characteristic equation can be solved for ω_n and ζ, and the motor velocity in response to a step voltage input (or any other form of input voltage) can be plotted.

Often, during system design and evaluation it has been useful to use the following approximation for the characteristic equation to quickly determine the expected response and appropriateness of a particular motor:

$$s^2 + \frac{(JR+BL)}{JL}s + \frac{(K_T K_V + BR)}{JL} \approx \left(s + \frac{R}{L}\right)\left(s + \frac{K_V K_T}{JR}\right) \tag{4.2}$$

(see Section 6.3 for analysis of this approximation).

In this expression, $\frac{R}{L} = \frac{1}{TC_E}$, where TC_E is the electrical time constant and $\frac{K_V K_T}{JR} = \frac{1}{TC_M}$, where TC_M is the mechanical time constant.

As shown in Section 6.3, this approximation is true if $TC_M > TC_E$ and if $\zeta > 1$ which is not always true and can lead to incorrect conclusions during initial rapid evaluation and comparisons of motor data, as shown by the following three examples.

4.2.1.1 Voltage Drive

In the following examples, the input voltage directly commands an output velocity with an error voltage created by subtracting the resulting BEMF from the command input.

Example 1

Motor Data $R = 15\ \Omega$ $K_T = 382$ g cm A^{-1} $J = 0.035$ g cm s^2
             $L = 6$ mH $K_V = 3.8 \times 10^{-2}$ V rad^{-1} s^{-1} $B = 0$ (no data available)

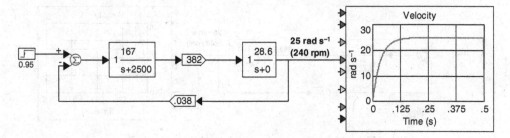

Figure 4.4 Step response; voltage drive, example 1

Exact: $s^2 + \dfrac{R}{L}s + \dfrac{K_T K_V}{JL} = s^2 + 2500s + 69\,124 = (s+28)(s+2472)$

$$\zeta = 4.7 \quad \omega_n = 263$$

Approx.: $TC_M = \dfrac{JR}{K_V K_T} = 36 \text{ ms} \rightarrow 27.7 \text{ rad s}^{-1}$

$$TC_E = \dfrac{L}{R} = 0.4 \text{ ms} \rightarrow 2500 \text{ rad s}^{-1}$$

Therefore the characteristic equation $= (s+27.7)(s+2500) = s^2 + 2528s + 69\,250$.

Use of the approximation produces the same result as the exact expression.

The voltage step response for this motor is shown in Figure 4.4 with the motor being commanded to 25 rad s^{-1} (240 rpm).

In this simulation, since $B = 0$ (no data available) the final torque error is zero.

Example 2

Motor Data $R = 2\,\Omega$ $K_T = 4032$ g cm A^{-1} $J = 8$ g cm s^2

                $L = 20$ mH $K_V = 0.40$ V rad^{-1} s^{-1} $B = 14.4$ g cm rad^{-1} s^{-1}

                                           (B measured with a retardation test)

Exact: $s^2 + \dfrac{(JR+BL)}{JL}s + \dfrac{(K_T K_V + BR)}{JL} = s^2 + 102s + 10\,363$

$$= (s+50+j88)(s+50-j88)$$

$$\zeta = 0.5 \quad \omega_n = 102$$

Approx.: $TC_M = \dfrac{JR}{K_V K_T} = 9 \text{ ms} \rightarrow 112 \text{ rad s}^{-1}$

$$TC_E = \dfrac{L}{R} = 10 \text{ ms} \rightarrow 100 \text{ rad s}^{-1}$$

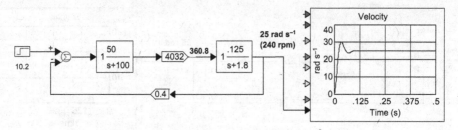

Figure 4.5 Step response; voltage drive, example 2

Therefore the characteristic equation $= (s + 112)(s + 100) = s^2 + 212s + 11\,200$

$$\zeta = 1 \quad \omega_n = 105$$

In this case, since the time constants are almost equal, the characteristic equation derived by using them leads to an incorrect value for the damping constant.

The voltage step response for this motor is shown in Figure 4.5.

Here the damping factor $B = 14.4$ g cm rad^{-1} s^{-1}, requiring a final steady-state torque of 360.8 g cm at 25 rad s^{-1}.

Example 3

Motor Data $R = 2.59\ \Omega$ $K_T = 3895$ g cm A^{-1} $J = 0.311$ g cm s^2
$L = 35.4$ mH $K_V = 0.382$ V rad^{-1} s^{-1} $B = 0.79$ g cm rad^{-1} s^{-1}

$$\text{Exact: } s^2 + \frac{(JR + BL)}{JL}s + \frac{(K_T K_V + BR)}{JL} = s^2 + 75.7s + 135\,333$$

$$= (s + 38 + j366)(s + 38 - j366)$$

$$\zeta = 0.1 \quad \omega_n = 368$$

$$\text{Approx.: } TC_M = \frac{JR}{K_V K_T} = 0.54 \text{ ms} \rightarrow 1847 \text{ rad s}^{-1}$$

$$TC_E = \frac{L}{R} = 14 \text{ ms} \rightarrow 73.2 \text{ rad s}^{-1}$$

Therefore the characteristic equation $= (s + 73.2)(s + 1847) = s^2 + 1920s + 135\,200$

$$\zeta = 2.6 \quad \omega_n = 368$$

In this case, the electrical time constant is actually larger than the mechanical time constant, which tends to be the case in brushless motors with small diameter permanent magnet rotors. Again, as in Example 2, the approximate calculation of the damping constant is incorrect.

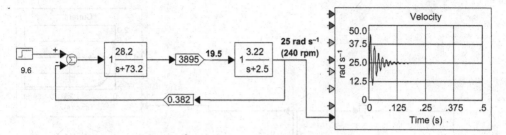

Figure 4.6   Step response; voltage drive, example 3

The voltage step response for this motor is shown in Figure 4.6.

Again a damping factor $B = 0.79$ g cm rad^{-1} s^{-1} requires a final steady-state error of 19.5 g cm at 25 rad s^{-1}.

4.2.1.2   Current Drive

Using current rather than voltage drive of the motor places the amplifier inside a loop by adding a second feedback path, as shown in Figure 4.7.

The transfer function now becomes:

$$\frac{R}{C} = \frac{\dfrac{FK_T}{JL}}{s^2 + \dfrac{[J(R + FA_I) + BL]}{JL}s + \dfrac{[K_T K_V + (R + FA_I)B]}{JL}} \tag{4.3}$$

There is now an "effective" resistance term, $(R_A + FA_I)$ in the second and third terms of the transfer function. Since B is usually small with respect to the other constants the result is that the third term will not appreciably change but the second term will increase in proportion to the relative values of R_A and FA_I. Since this term contains the damping factor ζ the result is that damping will increase, as shown in the following examples.

Note that if the current loop is opened ($A_I = 0$ and $F = 1$) this expression resolves to Equation 4.1.

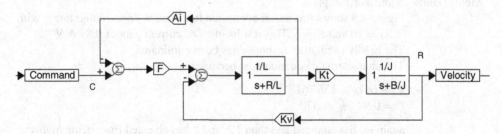

Figure 4.7   Current drive block diagram

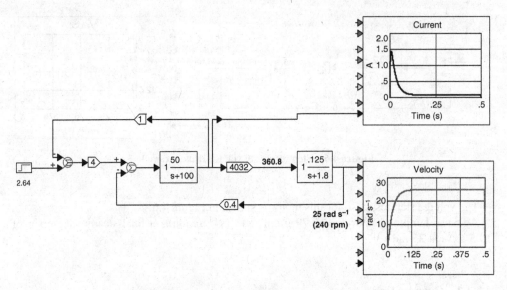

Figure 4.8   Step response; current drive, example 4

Also, the input voltage commands current, which in turn creates torque, resulting in acceleration of the load to a velocity limited by the damping constant.

Example 4

Motor Data: Same as Example 2
Figure 4.8 shows this motor connected in a current drive configuration with
$A_I = 1$ and $F = 4$. This results in a DC current gain of 0.67 A V^{-1}.
The characteristic equation now becomes:

$$s^2 + 302s + 10\,727$$
$$\zeta = 1.5 \quad \omega_n = 104$$

showing that ω_n has changed less than 2% and ζ has changed from being underdamped to somewhat overdamped

Example 5

Motor Data: Same as Example 3
Figure 4.9 shows this motor connected in a current drive configuration with
$A_I = 20$ and $F = 1$. This results in a DC current gain of 0.89 A V^{-1}.
The highly oscillatory response has been eliminated
The characteristic equation now becomes:

$$s^2 + 641s + 136\,761$$
$$\zeta = 0.9 \quad \omega_n = 370$$

again ω_n has changed less than 1% and ζ has changed from being highly underdamped to a damping factor of just less than 1.

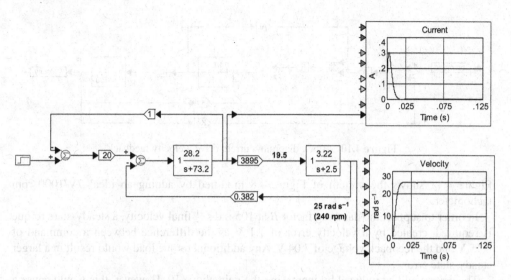

Figure 4.9   Step response; current drive, example 5

4.2.1.3   Current versus Voltage Drive

Since current drive appears to be so much better than voltage drive why would voltage drive ever be used?

If in Equation 4.3 it is assumed that $B = 0$ and $FA_I \to \infty$, then:

$$\frac{\theta'}{E_A} = \frac{K_T}{sJA_I} \tag{4.4}$$

which means that the velocity transfer function of a high gain current drive has a constant $-90°$ phase shift compared to a voltage drive which starts at a $0°$ phase shift at a frequency one octave below its first break point (typically determined by TC_M) and ends at $-180°$ phase shift, one octave above its second break point (typically determined by TC_E).

Therefore, low bandwidth (high inertia) systems could be candidates for voltage drive and be easier to stabilize than current drive, especially if the two break points are widely separated and the second occurs in close proximity to 0 db.

4.2.2   Velocity Loop

Figure 4.10 is the basic block diagram of a current driven motor with velocity feedback.

The closed loop transfer function is:

$$\frac{R}{C} = \frac{AFK_T/JL}{s^2 + \dfrac{[J(R + FA_I) + BL]}{JL}s + \dfrac{K_T(K_V + K_EFA) + (R + FA_I)B}{JL}} \tag{4.5}$$

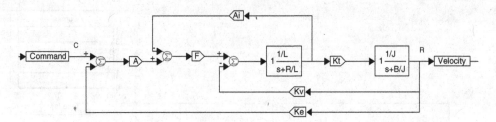

Figure 4.10   Block diagram; current drive, velocity feedback

Figure 4.11 shows the system of Figure 4.8 modified by adding an ideal 7V/1000 rpm tachometer.

In order to support the damping factor B at 105 rad s^{-1} final velocity, a steady-state torque is required, created by a velocity error of 1.1 V as the difference between a command of 8.15 V and the feedback voltage of 7.04 V. Any additional torque load would result in a larger steady-state error.

The error could be reduced by increasing the gain above 10. However, this would cause a larger initial overshoot and lead to instability.

Note also that a peak current of 32 A results due to the step nature of the command and final velocity is reached in less than 125 ms.

A better solution would be to use a ramp command or add PI compensation, as shown in Figure 4.12.

The PI compensation results in a virtual zero error and the input command and feedback signal are essentially equal. Final velocity is still achieved in 125 ms, but the peak accelerate current is now 5 A. The PI compensation essentially has two effects; it initially integrates the step command into a ramp, and in the steady-state forces the system to a virtual zero error while causing the error torque to support the load.

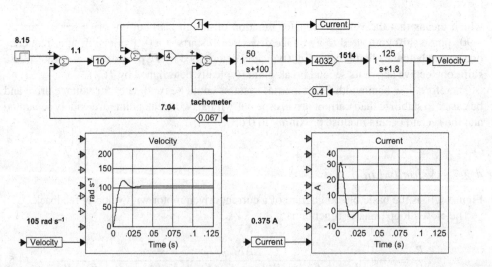

Figure 4.11   Step response; current drive, ideal tachometer (available in full color at www.wiley.com/go/moritz)

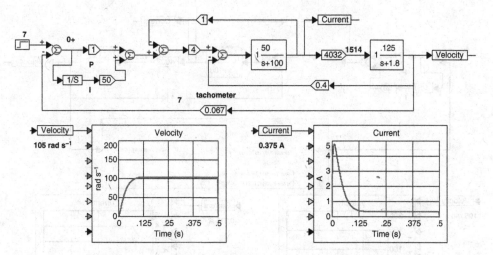

Figure 4.12   Step response; current drive, ideal tachometer, PI comp

Figure 4.13 shows the system reacting to a step load torque of 14 715 g cm at 5 s after start of motion. The system recovers within 100 ms and continues at 1000 rpm with zero error.

Figure 4.14 shows the ideal tachometer replaced with an emulated tachometer that has been scaled for 7 V/1000 rpm, with a 12 segment commutator and a ripple content of 3% RMS of the calibration. The ripple frequency of 12 kHz at 1000 rpm is well outside of the system bandwidth and has no effect on the output velocity.

When the velocity is reduced to 5.25 rad s^{-1}, as shown in Figure 4.15, the ripple frequency falls within the bandwidth and causes a 6% peak-to-peak variation of the output velocity.

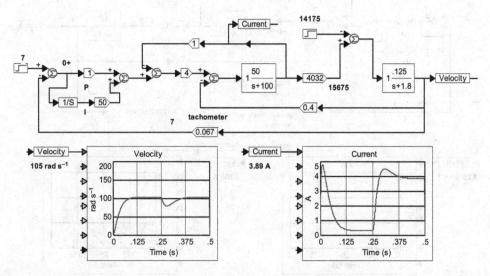

Figure 4.13   Same as Figure 4.12 with load torque step (available in full color at www.wiley.com/go/moritz)

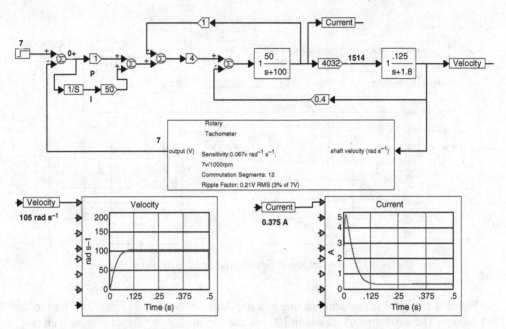

Figure 4.14 Step response; current drive, emulated tach., 1000 rpm (available in full color at www.wiley.com/go/moritz

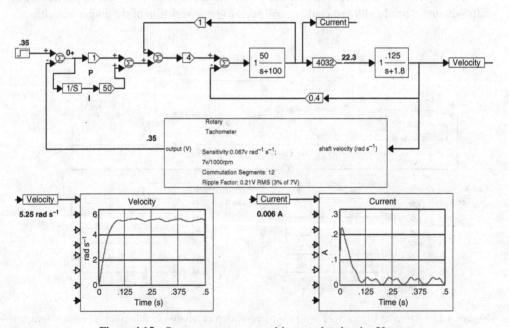

Figure 4.15 Step response; current drive, emulated tach., 50 rpm

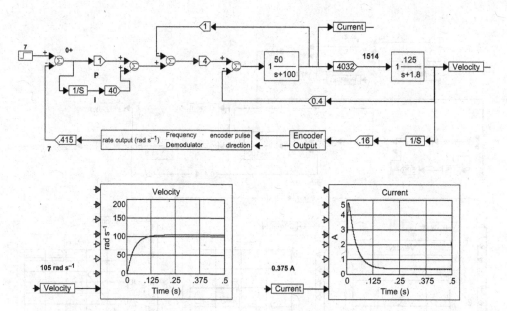

Figure 4.16 Step response, current drive, emulated encoder, 1000 rpm (available in full color at www.wiley.com/go/moritz)

In Figure 4.16 the emulated tachometer has been replaced with an emulated 2000 line encoder followed by a frequency demodulator to create a virtual velocity feedback signal.

The demodulator is of the trapezoidal integrator type followed by a 500 Hz output filter.

4.2.3 Position Loop

Figure 4.17 is the basic block diagram of a current driven motor with both velocity and position feedback.

The closed loop transfer function is:

$$\frac{R}{C} = \frac{BK_T FA/JL}{s^3 + \frac{[J(R+FA_I)+BL]}{JL}s^2 + \frac{K_T(K_V+K_EFA)+(R+FA_I)B}{JL}s + \frac{BK_PK_TFA}{JL}}$$

(4.6)

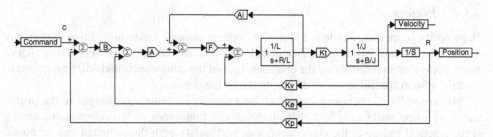

Figure 4.17 Block diagram; current drive, velocity and position feedback

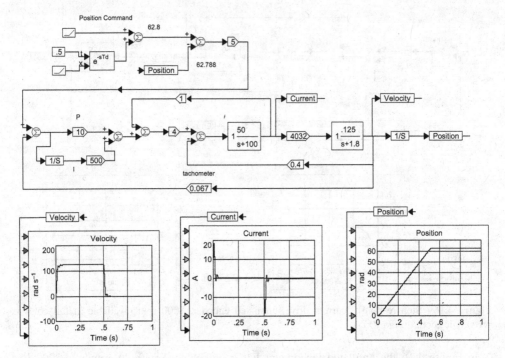

Figure 4.18   Same as Figure 4.12 with position feedback; 10 rev. command (available in full color at www.wiley.com/go/moritz)

Figure 4.18 shows Figure 4.12 with the position feedback path added and responding to a command to rotate 10 revolutions (62.8 rad) in 0.5 s.

The final position (62.788 rad) represents a 0.02% accuracy.

Be careful when interpreting simulation results. This simple model resulted in an accuracy of 0.032% when rotating one revolution but, for the 2000 count encoder being considered, this would mean 0.64 counts, which of course is impossible. A $\pm$ 1 count would represent 0.05% and a 2 count (0.1%) would be more realistic.

Figure 4.19 shows Figure 4.16 under the same conditions.

4.3   The Velocity Profile

4.3.1   Preface

The velocity, or motion, profile is a plot of the system velocity versus time. From the velocity profile, acceleration, jerk and distance can also be derived and displayed versus time. All of these profiles aid in determining the characteristics of the components that will be required to meet the system specifications as demonstrated by the profiles.

This process is a continuous "what if?" process. As it continues, changes in the profile may have to be made to reflect characteristics and limitations of the components, ending in the practical choice of the components that will satisfy both the technical and economic requirements of the final design.

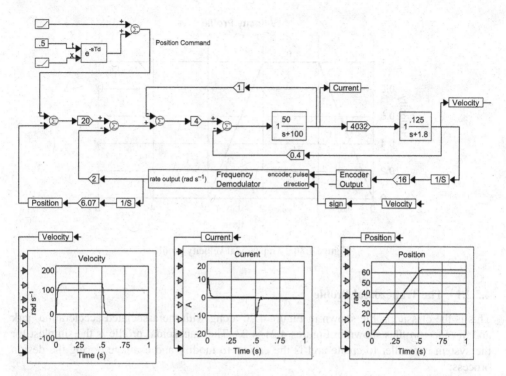

Figure 4.19 Same as Figure 4.16 with position feedback; 10 rev. command (available in full color at www.wiley.com/go/moritz)

The advent of the computer and microprocessor with associated software has made it possible to create and synthesize many different profiles which for many years were difficult or impossible to achieve. The profile can now even be modified "on the fly" in response to changes in system loads, forces or move requirements. Various profiles can be stored in memory and called up as needed for the system to perform selected motions.

Motion profiles can be divided into two general categories:

- Incremental Motion – Repetitive motion between two positions in which time and distance are of prime importance and velocity is secondary. Example: A pick and place operation in which an object is moved from "A" to "B" and the process is repeated "*n*" times
- Constant Motion – Motion in which velocity and distance are of prime importance. Example: A machining operation in which a certain velocity must be achieved as quickly as possible and maintained for a certain distance.

4.3.2 Incremental Motion

The following describes profiles used in most contemporary motion control systems.

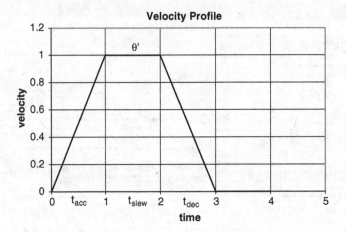

Figure 4.20   Trapezoidal velocity profile

4.3.2.1   The Trapezoidal Profile

This is the classic profile shown in Figure 4.20 along with the associated acceleration, jerk and distance profiles shown in Figures 4.21–4.23. The trapezoidal profile is the simplest for the system controller to create and is the easiest to modify and evaluate during the design process:

- Velocity

$$\theta'_{ACC}(t) = \frac{\theta'_M}{t_{ACC}}t \quad \theta'_{SLEW}(t) = \theta'_M \quad \theta'_{DEC}(t) = \frac{\theta'_M}{t_{DEC}}t \tag{4.7}$$

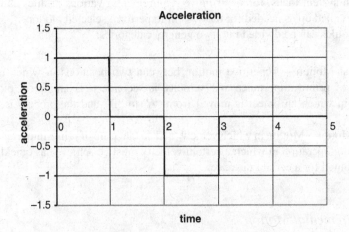

Figure 4.21   Trapezoidal profile: acceleration

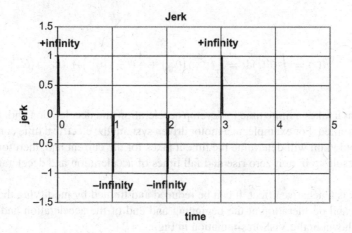

Figure 4.22 Trapezoidal profile: jerk

- Acceleration

$$\theta''_{ACC}(t) = \frac{\theta'_M}{t_{ACC}} \quad \theta''_{SLEW}(t) = 0 \quad \theta''_{DEC}(t) = \frac{\theta'_M}{t_{DEC}} \tag{4.8}$$

- Jerk

$$\theta'''(t) = \pm\text{infinity:} \tag{4.9}$$

Since jerk is the rate of change of acceleration and deceleration, in the ideal theoretical profile illustrated here, in which acceleration and deceleration change from zero to some constant value in zero time, the jerks are infinite, as shown.

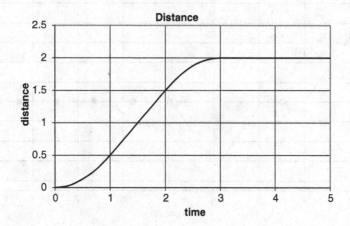

Figure 4.23 Trapezoidal profile:distance

● Distance.

$$\theta(t) = \int \theta'(t)\mathrm{d}t = \frac{1}{2}(t_{acc})\left(\theta'_M\right) + \frac{1}{2}(t_{dec})\left(\theta'_M\right) + (t_{slew})\left(\theta'_M\right) \qquad (4.10)$$

In any system it takes a finite time to develop acceleration and deceleration and, therefore, the jerk will be limited. For example, in a motor driven system, the electrical time constant and the amplifier bandwidth will determine the time it takes for the current and, therefore, the torque to change, resulting in non-zero rise and fall times of acceleration and deceleration, creating a finite jerk.

Since jerk is a negative effect, it can be reduced and limited by modifying the slope of the acceleration and deceleration at the beginning and end of the acceleration and deceleration periods, as shown in the VisSim simulation in Figure 4.24.

Note, if the move distance and time are constant, rounding off the acceleration and deceleration curves will cause an effective delay in the profile, resulting in an increase in the move time.

This can be compensated for by increasing the slope of the straight line portion of the curves and/or increasing the slew velocity.

If the slope is increased, the peak torque and, therefore, the peak motor current will be larger than that predicted by the ideal profile.

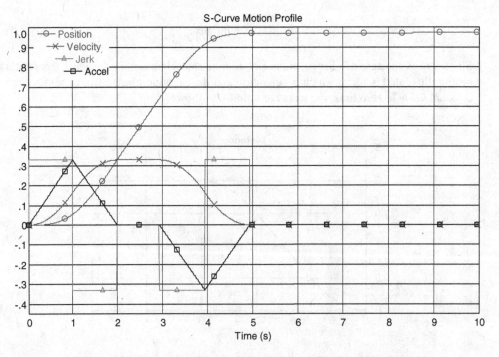

Figure 4.24   S-Curve motion profile: velocity, acceleration, jerk, distance

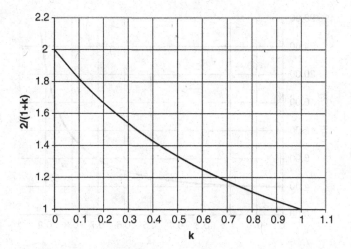

Figure 4.25 Normalized velocity; fixed move distance and time

If the slew velocity is increased, it might be necessary to select a motor with lower voltage constant than initially anticipated, resulting in a lower torque constant and an increased peak current, even if the linear portion of the curves is not changed.

It is interesting to develop the relationship between the slew (or peak) velocity, the three time periods and the total move distance.

$$\text{Assume } t_{acc} = t_{dec} \text{ and } t_{slew} = kT \quad \text{where} \quad 0 \leq k \leq 1$$

$$T = \text{total move time} = t_{acc} + t_{dec} + t_{slew}$$

$$\text{Then: } \theta'_M = \left(\frac{2}{1+k}\right)\left(\frac{\theta}{T}\right) \text{ and } \theta'' = \left(\frac{4}{1-k^2}\right)\left(\frac{\theta}{T^2}\right) \tag{4.11}$$

which states that if θ (the move distance) and T (the move time) are fixed, as determined by the system requirements, then if k equals 0 ($t_{\text{slew}} = 0$), the trapezoid becomes a triangle and the system experiences maximum peak velocity and minimum acceleration.

As k approaches 1, the peak velocity decreases and the acceleration approaches infinity, with the profile theoretically becoming a rectangle.

These two effects are shown graphically in Figures 4.25 and 4.26.

Note that in the range $0 \leq k \leq 0.5$ the peak velocity drops by 66.5% from the maximum (triangle) value while the acceleration increases only 33.3% above the minimum (triangle) value.

4.3.2.2 The "S" or Cosine Profile

As discussed in Section 4.3.2.1, the jerk effect can be controlled by rounding off the beginning and end of the acceleration and deceleration portions of the profiles, as shown in Figure 4.24.

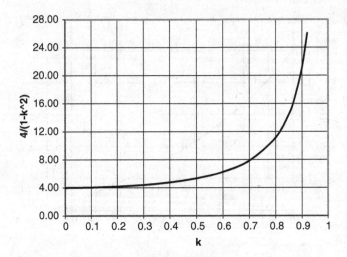

Figure 4.26 Normalized acceleration; fixed move distance and time

The ultimate rounding would be to use an easily implemented cosine profile, as shown in Figure 4.27, together with a triangular profile for comparison

- Velocity

$$\theta'(t) = \frac{\theta'_M}{2} - \frac{\theta'_M}{2} \cos \omega t = \frac{\theta'_M}{2}(1 - \cos 2\pi f t) = \frac{\theta'_M}{2}\left(1 - \cos \frac{\pi t}{t_{ACC}}\right) \quad (4.12)$$

$$f = \frac{1}{2t_{ACC}}$$

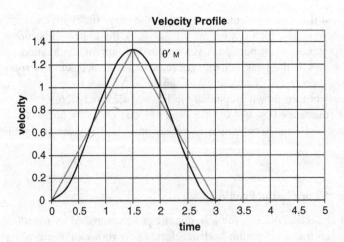

Figure 4.27 Cosine velocity profile

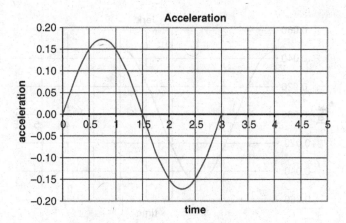

Figure 4.28   Cosine profile: acceleration

During the first half of the acceleration, less distance is covered than during the first half of the equivalent triangular profile. However, during the second half, more distance is covered compared to the triangular profile, with the result that the total distance is the same for both profiles, thereby not increasing the total time as is the case of just rounding the start and finish of the acceleration/deceleration portions of the trapezoidal profile. In order to achieve this the cosine profile has higher acceleration/deceleration at the midpoint compared to the constant value of the triangular profile when covering the same distance.

$$\text{For the trapezoid, the slope (the acc)} = \theta'' = \frac{\theta'_M}{t_{ACC}}$$

$$\text{For the cosine: slope} = \frac{\partial \theta'(t)}{\partial t} = \frac{\theta'_M}{2} \left(\frac{\pi}{t_{ACC}} \sin \frac{\pi t}{t_{ACC}} \right) \tag{4.13}$$

at $t = \frac{t_{ACC}}{2}$

$$\text{slope} = \frac{\pi \theta'_M}{2t_{ACC}} = 1.57 \left(\frac{\theta'_M}{t_{ACC}} \right) \tag{4.14}$$

The cosine requires 57% more peak torque than a trapezoid.
- Acceleration (Figure 4.28)

$$\theta''(t) = \frac{d\theta'(t)}{dt} = \frac{\theta'_M \pi}{2t_{ACC}} \sin \frac{\pi t}{t_{ACC}} \tag{4.15}$$

- Jerk (Figure 4.29)

$$\theta'''(t) = \frac{d\theta''(t)}{dt} = \frac{\theta'_M \pi^2}{2t_{ACC}^2} \cos \frac{\pi t}{t_{ACC}} \tag{4.16}$$

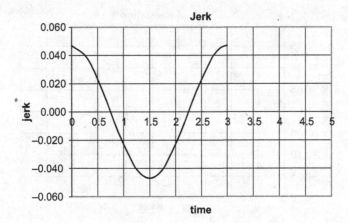

Figure 4.29 Cosine profile: jerk

Note that the jerk has a finite, predictable maximum value. It can be adjusted by varying the maximum velocity, the acceleration/deceleration, or both.

- Distance (Figure 4.30).

$$\theta(t) = \int \theta^t(t)dt = \frac{\theta'_M t_{ACC}}{2}\left(\frac{t}{t_{ACC}} - \frac{1}{\pi}\sin\frac{\pi t}{t_{ACC}}\right) \tag{4.17}$$

4.3.2.3 The Parabolic Profile

The parabolic profile is interesting in that it has been shown to result in the lowest motor dissipation for an incremental move, when compared to a trapezoidal or triangular move, for a constant distance and time (θ and t) [1]:

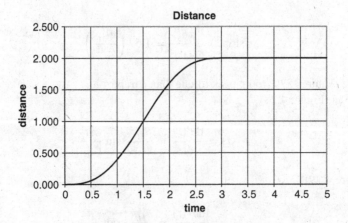

Figure 4.30 Cosine profile: distance

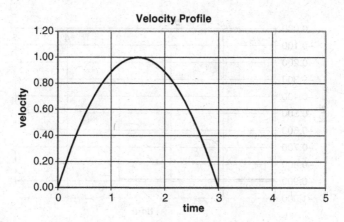

Figure 4.31   Parabolic velocity profile

Motor dissipation for a 1/3, 1/3, 1/3 trapezoidal move will be 12.5% higher than for a parabolic move.

Motor dissipation for a triangular move will be 33.3% higher than for a parabolic move.

For many years the parabolic profile was mainly of academic interest since it was not possible to create it in the system controller, but with the advent of modern control hardware and software the parabola can be considered.

The parabolic profile is shown in Figure 4.31.

- Velocity (Figure 4.32)

$$\theta'(t) = \frac{2\theta'_M}{t_{ACC}}\left(t - \frac{t^2}{2t_{ACC}}\right) \tag{4.18}$$

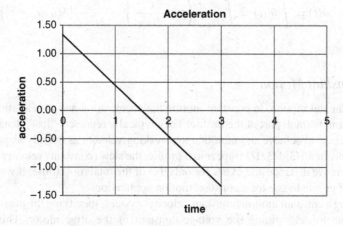

Figure 4.32   Parabolic profile: acceleration

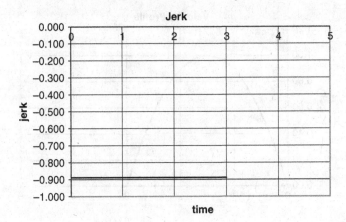

Figure 4.33 Parabolic profile: jerk

- Acceleration (Figure 4.33)

$$\theta''(t) = \frac{2\theta'_M}{t_{ACC}}\left(1 - \frac{t}{t_{ACC}}\right) \tag{4.19}$$

- Jerk

$$\theta'''(t) = -\frac{2\theta'_M}{t_{ACC}^2} \tag{4.20}$$

Jerk exhibits an infinite peak at the start and end of the move and a constant value during the move.
- Distance (Figure 4.34).

$$\theta(t) = \int \theta'(t) = \int \frac{2\theta_M}{t_{ACC}}\left(t - \frac{t^2}{2t_{ACC}}\right) = \frac{\theta'_M}{3t_{ACC}^2}\left(3t_{ACC}t^2 - t^3\right) \tag{4.21}$$

4.3.3   Constant Motion

Unlike incremental motion, in constant motion the acceleration and deceleration parts of the profile constitute a small part of the profile. They typically represent "lost" distance and time, since the object is to achieve and maintain the working velocity as quickly as possible.

For example, in a 1/3, 1/3, 1/3 trapezoidal profile, the slew (constant) velocity region is only $\frac{1}{2}$ of the total move distance and exists for only 1/3 of the total move time. It would not be the most efficient profile to use for a constant motion application.

In designing a constant motion (constant velocity) system, specifying the maximum velocity will contribute to determining the voltage constant of the drive motor. This in turn will determine the torque constant and, therefore, the peak current required during acceleration,

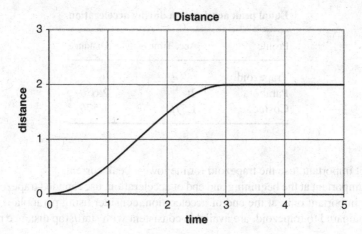

Figure 4.34   Parabolic profile: distance

deceleration and slew. The current required during slew will depend on the continuous torque loads; the current required during acceleration and deceleration will depend on the torque loads and the specific acceleration/deceleration profile to be used.

The profile to be used will depend on trade-offs between start/stop distance, time and acceleration (current).

It is interesting to compare the three profiles during the acceleration portion of a continuous motion profile with respect to equal distance, equal time and equal acceleration. The following shows these comparisons, using the trapezoid conditions as the base for the comparison.

Equal distance during acceleration

Profile	Acc. Time	Peak Acc.
Trapezoid	t	θ_T''
Parabola	$0.75t$	$2.67\theta_T''$
Cosine	t	$1.57\theta_T''$

Equal acceleration time during acceleration

Profile	Distance	Peak Acc.
Trapezoid	θ_T	θ_T''
Parabola	$\frac{4}{3}\theta_T$	$2\theta_T''$
Cosine	θ_T	$1.57\theta_T''$

Equal peak acceleration during acceleration

Profile	Acc. Time	Distance
Trapezoid	t	θ_T
Parabola	$2t$	$2.67\theta_T$
Cosine	$1.57t$	$1.57\theta_T$

Summary

If jerk is not important, use the trapezoid for the lowest peak current.

If jerk is important at the beginning and end of acceleration, use rounded trapezoid or cosine.

If jerk is important only at the end of acceleration, consider using parabola if higher peak currents, compared to trapezoid, are available consistent with start/stop distance requirement.

4.3.4   Profile Simulation

The three profiles being reviewed can easily be simulated using conventional simulation blocks, as shown in the following. In each case the simulation provides the velocity and distance commands that would be the inputs to a rotational system to move 20 rev in 4 s.

4.3.4.1   Trapezoid

Figure 4.35 shows the construction of a general purpose trapezoidal profile generator in which the inputs are velocity, acceleration, deceleration, acceleration time and run or slew time. The

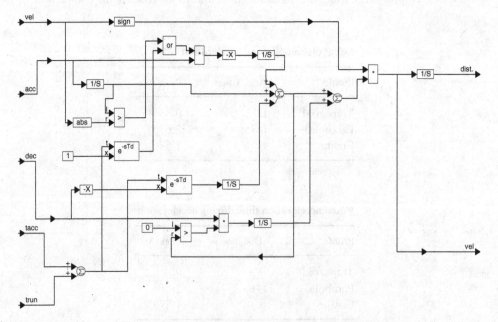

Figure 4.35   Programmable trapezoidal profile generator

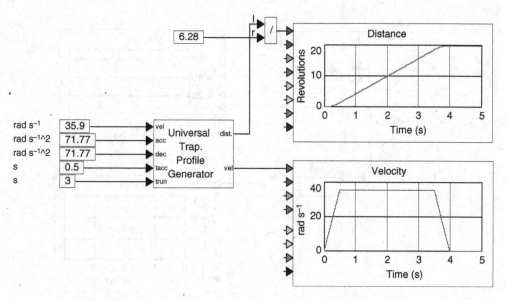

Figure 4.36 Profile generator developing profile for 20 revolutions in 4 s

outputs are velocity and distance. In use, the inputs can be modified as the simulation is run, observing the effect on the velocity and distance until satisfactory outputs (required distance within the allotted time at acceptable slew velocity) are achieved.

Figure 4.36 shows this profile generator being used to develop the profile for a rotation of 20 rev in 4 s. An arbitrary acceleration time of 0.5 s was selected and by choosing equal acceleration and deceleration values, the deceleration time also became 0.5 s. For this configuration, the run (slew) velocity required is 35.9 rad s^{-1}.

4.3.4.2 Cosine

Figure 4.37 shows a cosine profile generator, configured to also command 20 rev in 4 s. The peak velocity is 63 rad s^{-1}.

4.3.4.3 Parabola

Figure 4.38 shows a parabola profile generator, again generating commands for 20 rev in 4 s. Here the peak velocity is 47 rad s^{-1}, lower than the cosine profile but, as previously mentioned, the parabola involves higher acceleration and deceleration which these profiles graphically illustrate.

4.4 Feed Forward

A typical system that is stabilized by PID compensation can, depending on the PID settings required to achieve the required response and stability, be relatively slow acting in response to rapid changes in the command or changes in the load parameters.

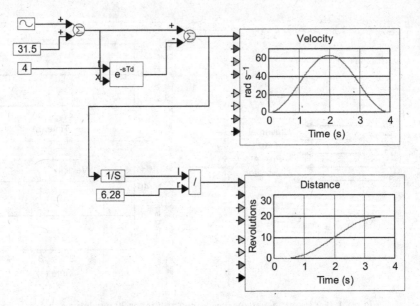

Figure 4.37   Cosine profile generator; 20 revolutions in 4 s

Feed forward control can be used outside of the normal closed loop to compensate for such changes.

Figure 4.39 shows the prototypical closed loop block diagram with a feed forward path added.

There are two summing junctions; the first compares the input reference R to the feedback signal CH to create the normal error signal E, the second compares the output of the

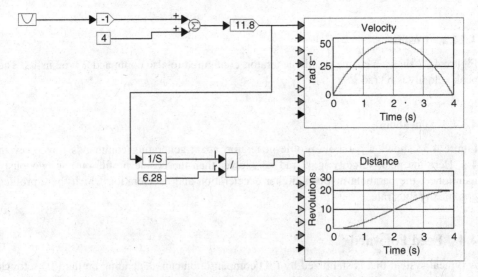

Figure 4.38   Parabola profile generator; 20 revolutions in 4 s

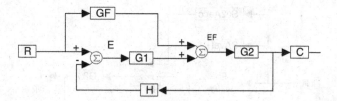

Figure 4.39 General closed loop block diagram with feed forward

compensating block EG_1 with the feed forward signal RG_F to create the final error signal to be amplified and used to power the motor and load.

Note that the feed forward signal completely bypasses G_1 which contains the normal filtering and compensation functions, such as PID, lead/lag, tachometer and so on. As such it has no direct impact on the stabilizing dynamics of the system and just contributes directly to the generation of torque or force.

Although Figure 4.39 shows the feed forward signal originating from R, it can actually come from any point or multiple points in the system which can affect performance and which can be measured and mathematically modeled. This requirement has made feed forward especially easy to implement with the advent of digital computer controllers in which such control algorithms can be created and limited only by the inventiveness and software capability of the designer.

Referring to Figure 4.39, the following transfer function can be derived:

$$\frac{C}{R} = \frac{G_F G_2}{1 + G_1 G_2 H} + \frac{G_1 G_2}{1 + G_1 G_2 H} \tag{4.22}$$

$$\text{If } G_F = 0 \qquad \frac{C}{R} = \frac{G_1 G_2}{1 + G_1 G_2 H} \tag{4.23}$$

which is the normal closed loop transfer function

$$\text{If } G_1 = 0 \qquad \frac{C}{R} = G_F G_2 \tag{4.24}$$

which is an open loop expression determined by the dynamics of G_F and G_2

Now, if $G_F = \frac{1}{G_2}$, then $C = R$ and theoretically a perfect control situation results.

Since G_2 consists mainly of the motor/load transfer function, it will have the form:

$$\frac{A}{s^2 B + sC} \tag{4.25}$$

$$\text{resulting in } G_F = s^2 \frac{B}{A} + s \frac{C}{A} \tag{4.26}$$

meaning that the feed forward is created by processing the acceleration and velocity of the input position reference signal.

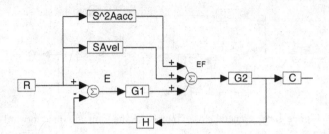

Figure 4.40   Closed loop block diagram with velocity and acceleration feed forward

Actually, G_1 and G_F both exist in any realizable system and the block diagram becomes that shown in Figure 4.40, which will allow independent adjustment of acceleration (Aacc)-and velocity (Avel)-dependent feed forward in addition to the PID terms contained in G_1.

Figure 4.13 (see Section 4.2) shows a velocity control system that has been adjusted for no overshoot and subjected to a disturbance torque when at speed, resulting in a 23% peak dip in velocity before recovering to 1000 RPM.

Figure 4.41 shows this same system modified to include a feed forward signal developed as a function of the disturbance torque, resulting in only a 5% peak dip in velocity.

Figures 4.42 and 4.43 show this same system modified to PID compensation and operating in a position mode, responding to a trapezoidal velocity command.

A disturbance torque is applied at 1.5 s, once the system has settled to a steady state after rotating 10 rev.

In Figure 4.42 the feed forward path is not connected. In Equation 4.26 the feed forward path is connected. Compare the current and PID error waveforms. The current required to

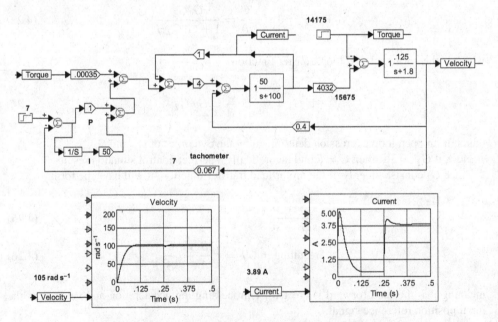

Figure 4.41   System of Figure 4.13 with feed forward added

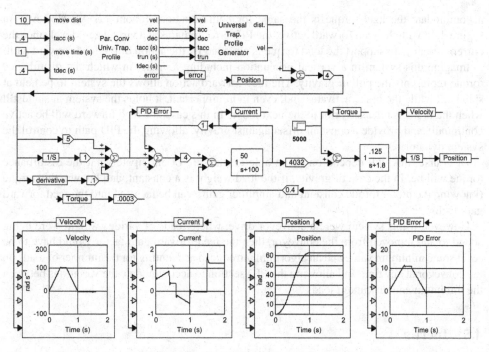

Figure 4.42   System of Figure 4.13 with PID in position mode; no feed forward

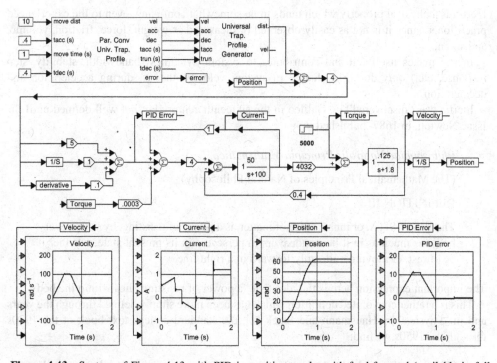

Figure 4.43   System of Figure 4.13 with PID in position mode; with feed forward (available in full color at www.wiley.com/go/moritz)

accommodate the load torque is the same for both situations (about 1.3 A). However, in Figure 4.43 with the feed forward active, the PID error remains at a virtual zero level since the current required to support the load torque is obtained by the action of the feed forward path.

Imagine this system in a vertical orientation including a brake, in which the disturbance torque represents the pull of gravity. The feed forward action allows the system to be held at standstill while the brake activates and, even more important, it holds the system at standstill when the brake is released prior to the next move. In this case, the feed forward will be active continually and provide a constant offset against gravity, allowing the PID path to control the system dynamics.

Of course, this will depend on being able to measure or anticipate what the disturbance torque will be. In the case of gravity, if the load weight is a constant, and known, this torque (knowing the motor torque constant and amplifier gain), can be factored into the feed forward gain factor.

However, if this system were a conveyor drive, where loads of various weight were being added to and removed from the conveyor, then the feed forward gain factor would have to be set at some minimum value and an algorithm introduced accounting for the number of loads on the conveyor at any time and allowing the PID section to accommodate the variations between the minimum and maximum loads.

4.5 Inertia

4.5.1 Preface

Inertia is a physical property which tends to be somewhat confusing, even to the experienced practitioner, since it is not as easily observed or measured as weight, force, friction, volume, and so on.

Inertia makes itself felt and contributes to system dynamics (bandwidth, stability, step response, etc.) *only* during a change in system velocity, that is, during acceleration and deceleration.

Inertia was "discovered" by Galileo in the sixteenth century but not well defined until Sir Issac Newton, in 1687, published in his

Philosophiae Naturalis Principia Mathematica

(The Mathematical Principles of Natural Philosophy)

DEFINITION III

The VIS INSITA, or innate force of matter, is a power of resisting, by which every body, as much as in it lies, endeavors to persevere in its present state, whether it be of rest, or moving uniformly forward in a right line.

The important expression here is the idea that "a power of resisting" lies within the body.

This is a rather convoluted definition which has been more simply defined through the years and in "Physics" by Hausmann and Slack [2], the standard college text book of the 1930s through the 1950s we find:

"The opposition which a body offers to any change of motion, whereby an unbalanced force is needed to give it linear acceleration, is known as inertia"

"Mass may be considered as that property of an object by virtue of which it possesses inertia"

Inertia makes itself felt in both linear (translation) and rotary (rotation) motion according to:

$F = mS''$ translation (F = force, m = mass, S'' acceleration)

$T = J\theta''$ rotation (T = torque, J = inertia moment, $= \theta''$ acceleration)

The important difference between these two effects, in applying them in motion system design, is that m is dependent only on the volume and density of the body, whereas J is dependent on the volume, density and physical configuration of the body.

For example, assume we have two cylinders with radii and lengths as shown in Figure 4.44, made of the same material with density ρ.

Since both cylinders, 1 and 2, have the same volume, they will have the same weight (W) and therefore in translation will require the same accelerating force (F):

$$W = \rho\pi R^2 L \therefore W_1 = \rho\pi \left(\sqrt{2}r\right)^2 l = \rho\pi 2r^2 l \quad W_2 = \rho\pi 2r^2 l \qquad (4.27)$$

$$F = \frac{W}{g}S'' \qquad (4.28)$$

However, in rotation, the inertia J of a cylinder is:

$$J = \frac{W}{2g}R^2 \qquad (4.29)$$

$$\text{Therefore for cylinder 1: } J = \frac{W}{2g}\left(\sqrt{2}r\right)^2 = \frac{W}{g}r^2 \qquad (4.30)$$

$$\text{Whereas, for cylinder 2: } J = \frac{W}{2g}r^2 \qquad (4.31)$$

Since cylinder 1 has twice the inertia of cylinder 2, it will require twice the torque to achieve the same acceleration.

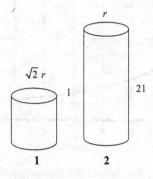

Figure 4.44   Cylinder comparison; equal weight, unequal inertia

4.5.2   Motor Selection

In choosing and evaluating a motor, a review of various manufacturers' data sheets is taken in order to make the selection.

Quite often, a data sheet will list a "maximum acceleration capability" for the motor.

This is based on the peak torque rating, which is usually available for only 1–3 s to prevent motor overheating. However, what is sometimes not directly stated, or stated in very fine print at the bottom of the data sheet, is that this rating is only for the motor in an unloaded condition.

If a plot is made of acceleration versus load inertia normalized to motor inertia the true capability of any motor can be shown.

$$\text{Let } a = J_{LOAD}/J_{MOTOR} \quad \text{where} \quad 0 \le a \le \infty \tag{4.32}$$

$$\text{and } \theta''_M = \frac{T_{ACC}}{J_{MOTOR} + aJ_{MOTOR}} = \frac{T_{ACC}}{J_{MOTOR}(1+a)}; \theta''_{MAX} = \frac{T_{ACC}}{J_{MOTOR}} \quad \text{when } a = 0 \tag{4.33}$$

$$\text{then } \frac{\theta''_{MOTOR}}{\theta''_{MAX}} = \frac{1}{1+a} \tag{4.34}$$

A plot of $\dfrac{1}{1+a}$ versus a $\left(\dfrac{\theta''_{MOTOR}}{\theta''_{MAX}} \text{ vs. } \dfrac{J_{LOAD}}{J_{MOTOR}}\right)$ is shown in Figure 4.45.

Although rather obvious, this plot demonstrates dramatically how careful one must be in interpreting data sheets. It shows how rapidly acceleration decreases as load inertia increases. When load inertia is only five times motor inertia, the available acceleration is less than 20% of the so-called "maximum acceleration".

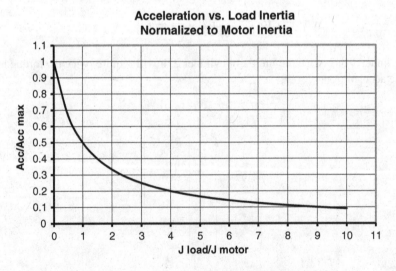

Figure 4.45   Normalized acceleration versus load inertia

4.5.3   Reflected Inertia – Gearhead

In general, a load inertia reflected to the input of a gearhead is:

$$J_{IN} = J_L/N^2 \tag{4.35}$$

also,

$$J_T = J_M + J_L/N^2 \tag{4.36}$$

and

$$\theta_M'' = N\theta_L'' \tag{4.37}$$

therefore,

$$T_{ACC} = J_T\theta_M'' = \left(J_M + J_L/N^2\right)N\theta_L'' = (J_MN + J_L/N)\theta_L'' \tag{4.38}$$

Taking the derivative of T_{ACC} with respect to N, equating to 0 and solving for N

$$\frac{\partial T_{ACC}}{\partial N} = \left(J_M - J_L/N^2\right)\theta_L'' = 0 \tag{4.39}$$

$$N = \sqrt{J_L/J_M} = N_{OPT} \tag{4.40}$$

This results in the total inertia $J_T = 2J_M$, or, in other words, minimum torque is achieved in a geared assembly when the reflected load inertia is equal to the motor inertia.

Although this effect is interesting and should be considered when initiating a design, it can very rarely be achieved for a number of reasons three of which are:

1. The term J_M must not only include the motor inertia itself but also additional components such as an encoder, a coupling and a brake rotor, each of which requires acceleration energy not then available to the load.
2. Even if these components are not present, the gearhead has inertia, as seen at its input shaft, and also represents a loss factor.
3. Off-the-shelf gearheads are only available in a finite number of whole number ratios, such as 3, 5, 7, 10, 15, and so on, which in the majority of applications will not provide a theoretical optimum match.

In the following sections the topics of torque and peak power versus the optimum condition are reviewed including an example describing conditions for using less than the optimum.

4.5.4 Torque versus Optimum Ratio – Gearhead

Define $T_{M(opt)}$ as the torque required under the condition in which the load inertia, T_L, reflected to the motor is equal to the motor inertia.

Then $T_{M(opt)} = 2J_M N \theta_L'' = 2\theta_L'' J_M \sqrt{J_L/J_M} = 2\theta_L'' \sqrt{J_L J_M}$ (4.41)

In general $T_M = \left(J_M + J_L/N^2\right) N\theta_L'' = J_M N\theta_L'' + J_L \theta_L''/N$ (4.42)

Let $b = N/N_{OPT}$ or $N = bN_{OPT}$ where $0 \le b \le \infty$ (4.43)

Then $T_M = J_M bN_{OPT}\theta_L'' + J_L\theta_L''/bN_{OPT}$ which can be reduced to   (4.44)

$T_M = (b + 1/b)\left(T_{M(opt)}/2\right)$ or $\dfrac{T_M}{T_{M(opt)}} = \dfrac{(b + 1/b)}{2}$ (4.45)

A plot of $\dfrac{(b+1/b)}{2}$ versus b $\left(\dfrac{T_M}{T_{M(opt)}} \text{ vs. } \dfrac{N}{N_{OPT}}\right)$ is shown in Figure 4.46.

Note that a mismatch of 0.5 to 2 ($\pm 100\%$) only requires 25% more torque than the ideal condition.

4.5.5 Power versus Optimum Ratio – Gearhead

Although the effect of reflected inertia through a gearhead on the torque, and therefore motor current, is important, it is also interesting to evaluate the effect on the peak power requirement during acceleration. If a trapezoidal profile is assumed during which the torque is a constant,

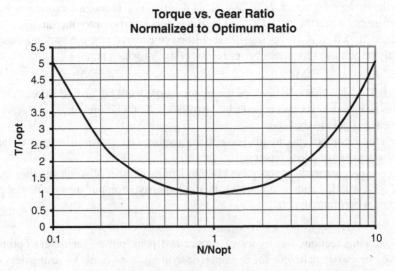

Figure 4.46   Normalized torque versus gear ratio

peak power will be occur at the end of the acceleration period.

$$\text{In general: Power} = \text{Torque} \times \text{Velocity} = T\theta' \tag{4.46}$$

$$\therefore \text{ Peak motor power } P_M = (J_M + J_L/N^2)(N\theta_L'')(N\theta_{LP}') \tag{4.47}$$

where $(J_M + J_L/N^2)(N\theta_L'')$ is the torque during acceleration and $N\theta_{LP}'$ is the maximum velocity.

$$\text{Then } P_M = J_M N^2 \theta_L'' \theta_{LP}' + J_L \theta_L'' \theta_{LP}' \tag{4.48}$$

$$\text{If } N^2 = N_{OPT}^2 \tag{4.49}$$

$$\text{then } P_{M(opt)} = (J_M)(J_L/J_M)\theta_L' \theta_{LP}' + J_L \theta_L'' \theta_{LP}' \tag{4.50}$$

$$= 2J_L \theta_L'' \theta_{LP}'$$

$$\text{Let } c = N/N_{OPT} \quad \text{or} \quad N^2 = c^2 N_{OPT}^2 \tag{4.51}$$

$$\text{then } P_M = J_M c^2 N_{OPT}^2 \theta_L'' \theta_{LP}' + J_L \theta_L'' \theta_{LP}' = c^2 J_L \theta_L'' \theta_{LP}' + J_L \theta_L'' \theta_{LP}' = (c^2 + 1)J_L \theta_L'' \theta_{LP}' \tag{4.52}$$

$$\text{therefore } \frac{P_M}{P_{M(opt)}} = \frac{(c^2 + 1)}{2} \tag{4.53}$$

A plot of $\dfrac{(c^2 + 1)}{2}$ versus $c \left(\dfrac{P_M}{P_{M(opt)}} \text{ vs. } \dfrac{N}{N_{OPT}} \right)$ is shown in Figure 4.47.

Note that having an optimum gearhead ratio ($N/N_{OPT} = 1$) which creates minimum torque does not result in minimum power. Minimum power results if the motor inertia is small with respect to the load inertia (N small) or conversely if it is assumed that the motor inertia is,

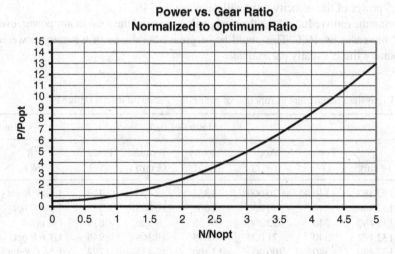

Figure 4.47 Normalized power versus gear ratio

theoretically, approaching zero, at which point only the load inertia is present and the power approaches one-half the optimal match power.

Example

Assume a system has a load with the following specifications:

For the first calculations, the gearhead inertia will be ignored.

$$J_L = 66.2 \text{ g cm s}^2 \quad \theta_L' = 250 \text{ rpm} = 26.2 \text{ rad s}^{-1}$$
$$t_{ACC} = 0.013 \text{ s}$$

A motor is selected with an inertia of: $J_M = 0.094 \text{ g cm s}^2$

For the first calculations, the gearhead inertia will be ignored.

$$\text{At the load: } \theta_L'' = 26.2/0.013 = 2000 \text{ rad sec}^{-2}$$

$$T_{ACC} = (66.2)(2000) = 132\,400 \text{ g cm}$$

$$\text{Peak load power} = \frac{132\,400 \times 250}{13\,824 \times 5250} \times 746 = 340 \text{ W}$$

$$\text{Optimal gearhead ratio} = N_{OPT} = \sqrt{66.2/0.094} = 26.6 : 1$$

$$\therefore J_T = 0.094 + 0.094 = 0.188 \text{ g cm s}^2$$

Table 4.1 shows how various parameters in this system will vary as the ratio is changed above and below the optimal value.

Notice in Table 4.2 that the optimum gearhead ratio results in the lowest I^2R loss. This is achieved at a motor velocity of 6650 rpm. However, by lowering the ratio from 26.6 to 20, the motor velocity decreases to 5000 rpm, a 25% decrease and the I^2R increases only 10%. This will result in increased bearing life and lower audible noise for both the motor and the gearhead. In addition, motor velocity-dependent losses (core losses), which are proportional to the 1.5 power of the velocity ratio will decrease by 54%.

Lowering the ratio reduces both the total peak power and the total motor power, even though the I^2R increases by 10%. This could be of great importance in a battery powered system where battery life is usually paramount.

Table 4.1 System parameters as function of gear ratio, gearhead inertia omitted

	Load		Motor + Load				Motor	
N	T_{acc} (g cm)	Power (W)	Acc. (r s^{-2})	Velocity (rpm)	Torque (g cm)	Power (W)	I^2R (W)	Power (W)
13.3	132 480	340	26 600	3 325	12 456	426	1.56 I^2R opt	86
20	132 480	340	40 000	5 000	10 368	533	1.1 I^2R opt	193
26.6	132 480	340	53 200	6 650	9 936	679	I^2R opt	339
35.5	132 480	340	71 000	8 875	10 368	945	1.1 I^2R opt	605
53.2	132 480	340	106 400	13 300	12 456	1702	1.56 I^2R opt	1362

Table 4.2 System parameters as function of gear ratio; gearhead inertia included

	Load		Motor + Load				Motor	
N	T_{acc} (g cm)	Power (W)	Acc. (r s^{-2})	Velocity (rpm)	Torque (g cm)	Power (W)	I^2R (W)	Power (W)
10	132 480	340	20 000	2 500	16 560	425	1.56 I^2R opt	85
15	132 480	340	30 000	3 750	13 824	533	1.1 I^2R opt	193
20	132 480	340	40 000	5 000	13 248	681	I^2R opt	341
26.7	132 480	340	53 333	6 667	13 824	947	1.1 I^2R opt	607
40	132 480	340	80 000	10 000	16 560	1702	1.56 I^2R opt	1362

Now, include the gearhead inertia as part of the motor when calculating the optimum ratio.

$$J_{GH} = 0.072 \text{ g cm s}^2 \quad J_{GH} + J_M = 0.166 \text{ g cm s}^2$$

$$\text{Optimal gearhead ratio} = N_{OPT} = \sqrt{66.2/0.166} = 20:1$$

$$\therefore J_T = 0.166 + 0.166 = 0.332 \text{ g cm s}^2$$

Table 4.2 now shows the effect on the system parameters of including the gearhead inertia Comparing these two tables shows the following:

1. The *real* optimum torque (13 248) is 1.33 times larger than the *theoretical* torque (9936).
2. The *real* motor velocity (5000) will be 1.33 times less than the *theoretical* motor velocity (6650), resulting in the total power remaining the same.
3. The *real* I^2R will be 1.76 times larger than the *theoretical* I^2R.
4. If it is desired to reduce the velocity further to 3750 rpm, then the *real* torque (13 824) increases to 1.39 times the *theoretical* optimum torque (9936), resulting in the *real* I^2R becoming 1.94 times larger than the *theoretical* I^2R.

Next, follow through on item 4. and assume the peak velocity of 3750 rpm should correspond to an optimum ratio, then:

$$N = \frac{3750}{250} = 15:1$$

therefore:

$$J_M = \frac{J_L}{N^2} - J_{GH} = 0.222 \text{ g cm s}^2$$

This means selecting a motor with 2.4 times the inertia of the initial selection, which will be a larger motor, but will have lower I^2R loss.

Also,

$$J_T = \frac{66.2}{15^2} + 0.222 + 0.072 = 0.588 \text{ g cm s}^2$$

which is 1.8 times the total inertia of the initial selection:

> Question: Which motor should be selected for the application?
>
> The smaller motor operating at somewhat less than optimum or the larger motor operating at optimum?
>
> The answer must include consideration of size, cost, weight, allowable peak speed, dissipation and ambient temperature and availability.

This example shows that care must be taken when selecting gearhead ratios to match load to motor inertia. A compromise will have to be selected between motor velocity, motor torque and motor I^2R. In general, for long life and low audible noise a low ratio is called for. However, a low ratio may lead to a relatively high torque and high I^2R dissipation.

In some cases the compromise might require the use of a motor larger than initially selected for evaluation.

4.5.6   Optimal Conditions

In the foregoing, optimal is defined as having the inertia reflected to the motor shaft equal to the motor inertia.

However, there can be other conditions that may be defined as optimal:

- Minimizing the total peak power could be an optimizing goal. As shown in Tables 4.1 and 4.2 this can be achieved by using a less than 1:1 inertia ratio, but at the cost of increased motor dissipation.
- Direct drive designs will not minimize the inertia ratio but will optimally provide a "stiff" mechanical assembly [3].
- A large motor inertia for systems not requiring rapid dynamics will optimally compensate for large load disturbances or eliminate the need for a gearhead entirely. One major manufacturer has introduced a new motor line by adding a flywheel to its standard line to increase the motor inertia by 3 to 10 times [4].
- Systems which have to operate with a range of load inertias, such as machining or robotics, even if they use a gearhead, belt and pulley, and so on., cannot have an optimum ratio as defined above. Their optimization will be related to the ability of the controller to provide efficient performance.

Systems with load to motor inertia ratios as high as 1000 : 1 have been successfully designed [5].

An interesting experiment was conducted to test the ability for a motor/controller to handle a load range of 0 to 50 times the motor inertia.

Motor: Model BM090-1A1A1 with:

$$K_T = 3319 \text{ g cm A}^{-1} \quad K_E = 0.326 \text{ V rad}^{-1} \text{ s}^{-1}$$
$$R_{TT} = 2.5 \text{ W} \quad L_{TT} = 4.5 \text{ mH}$$

$$J_M = 0.56 \text{ g cm s}^2$$

$$T_{cont} = 18\,940 \text{ g cm} \quad I_{cont} = 5.7 \text{ A}$$

$$T_{peak} = 56\,800 \text{ g cm} \quad I_{peak} = 17.1 \text{ A}$$

Amplifier/Controller: BDC6000

$$E_{\text{bus}} = 160 \text{ V} \quad I = 10 \text{ A cont./20 A peak}$$

Tachometer: Analog; 7 V/1000 rpm

A series of inertial load discs were fabricated to allow testing to be performed with loads equal to 2, 5, 10, 20 and 50 times the motor inertia. In each test, it was possible to adjust gain and compensation values to provide stable operation in both velocity and positional modes.

Figures 4.48 and 4.49 show simulations of the velocity test for the 0 and 50 times tests.

The amplifier section has a gain of 2 A V^{-1}, with the input limited to ± 10 V to limit the maximum current to ± 20 A. The output voltage is limited to ± 160 V.

Note that the 50 times inertia test simply required the loop gain to be 50 times higher than the value for the motor alone, showing how the system can be gain compensated for load inertia variations.

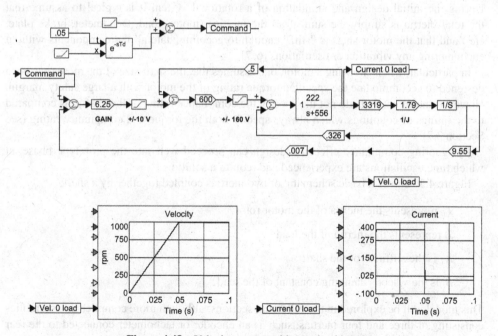

Figure 4.48   Velocity system with zero load inertia

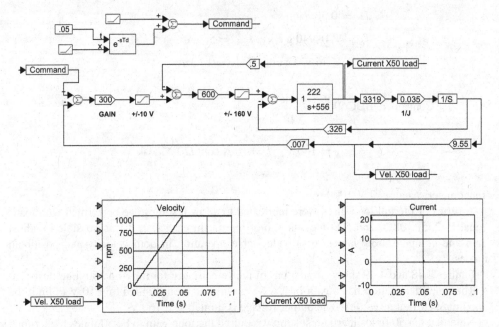

Figure 4.49   Velocity system with load inertia equal to 50 times motor inertia

4.6   Shaft Compliance

During the initial design and simulation of a rotational system it is typical to assume that the total inertia is simply the sum of all the inertias (motor, load, tachometer, brake plate, etc.) and that the motor shaft is "stiff" enough to accommodate all the components without encountering any vibration or oscillations [6, 7].

In particular, when choosing a motor, one assumes that the shaft size of the motor has been designed to accommodate the specified torque rating of the motor with a large safety margin. However, motor catalogs do not the specify the compliance rating of the shafts, as compared to the ratings of couplings which always specify both the torque and compliance rating (see Section 3.7).

By ignoring compliance effects, a design can proceed well into the prototype phase, at which time oscillations are experienced and require a solution.

Figure 4.50 shows a basic schematic of two inertias coupled together by a shaft:

J_m represents the inertia of the motor rotor

J_l represents the inertia of the load

K_l is the stiffness of the shaft

B_l is the viscous damping constant of the load

This model will be explored in the following sections, although more complicated assemblies consisting of three and four inertias, such as an encoder or tachometer connected to the rear of the motor with the load connected to the front could be simulated.

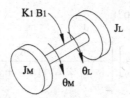

Figure 4.50   Two inertias coupled with shaft

4.6.1   Basic Equations

The equations expressing the torque relations for this model are:

$$T_m = J_m \theta_m'' + B_m \theta_m' + J_l \theta_l'' \tag{4.54}$$

$$J_l \theta_l'' = K_s \left(\theta_m - \theta_l\right) + B_l \left(\theta_m' - \theta_l'\right) \tag{4.55}$$

$J_l \theta_l''$ is that portion of T_m applied to J_l through the shaft.

$$\therefore T_m = (s^2 J_m + s B_m)\theta_m + s^2 J_l \theta_l \tag{4.56}$$

$$s^2 J_l \theta_l = (s B_l + K_s)\,\theta_m - (s B_l + K_s)\,\theta_l \tag{4.57}$$

Equations 4.56 and 4.57 can each be solved for θ_m, set equal to each other and solved for θ_l':

$$\theta_l' = \left[\frac{(s B_l + K_s)}{\left(s^2 J_l + s B_l + K_s\right)(s J_m + B_m) + s J_l \left(s B_l + K_s\right)} \right] T_m \tag{4.58}$$

Since B_l is typically unknown and difficult to measure, assume it to be 0 such that:

$$\theta_l' = \left[\frac{(K_s)}{\left(s^2 J_l + K_s\right)(s J_m + B_m) + s J_l K_s} \right] T_m \tag{4.59}$$

Equation 4.56 can be rewritten as:

$$\therefore T_m = s^2 J_m \theta_m + s B_m \theta_m + s^2 J_l \theta_l \tag{4.60}$$

since $s \theta_m = \theta_m'$

$$\therefore T_m = s J_m \theta_m' + B_m \theta_m' + s^2 J_l \theta_l \tag{4.61}$$

solving for θ_m'

$$\theta_m' = \frac{T_m}{s J_m + B_m} - \frac{s J_l \theta_l'}{s J_m + B_m} \tag{4.62}$$

Combining Equations 4.59 and 4.62:

$$\theta'_m = \frac{T_m}{sJ_m + B_m} - \left[\frac{sJ_l}{sJ_m + B_m}\right]\left[\frac{K_s}{(s^2J_l + K_s)(sJ_m + B_m) + sJ_lK_s}\right]T_m \qquad (4.63)$$

$$\theta'_m = \frac{T_m}{sJ_m + B_m} - \left[\frac{sJ_l}{sJ_m + B_m}\right]\left[\frac{\frac{K_s}{J_lJ_m}}{s^3 + \frac{B_m}{J_m}s^2 + \frac{(J_l + J_m)K_s}{J_lJ_m}s + \frac{B_mK_s}{J_lJ_m}}\right]T_m \qquad (4.64)$$

This is the equation which will be used to synthesize the effect of compliance as shown in the following sectionss.

4.6.2   System Components

Motor:	$R = 2.75\ \Omega$	$L = 13.7\ \text{mH}$

$J_m = 0.115\ \text{g cm s}^2$
$K_t = 3240\ \text{g cm A}^{-1}$
$K_e = 0.318\ \text{V rad}^{-1}\ \text{s}^{-1}$
$T_{cont} = 6800\ \text{g cm}$ $T_{peak} = 20\,200\ \text{g cm}$
$B_m = 0.338\ \text{g cm rad}^{-1}\ \text{s}^{-1}$ (measured per Section 2.6.1)
$K_s = 2677\ \text{N m rad}^{-1} = 2.73 \times 10^7\ \text{g cm rad}^{-1}$
(calculated for a steel shaft, 1.27 cm dia × 7.62 cm long; per 3.7/5)

Tachometer: $K_T = 2.5\ \text{V/1000rpm} = 0.024\ \text{V rad}^{-1}\ \text{s}^{-1}$
Load Disc $= J_l = 5 \times J_m = 0.575\ \text{g cm s}^2$
Amplifier: Current Type   Gain $= 1.7\ \text{A V}^{-1}$
Output voltage limit $= \pm160\ \text{V}$

4.6.3   Initial Simulation – Lumped Inertia

Figure 4.51 shows a simulation of the system, with the motor and load inertia as a single value in a closed loop velocity servo, responding to a step command to 100 rad s^{-1}.

The motor/load velocity shows a good response, with a 20% overshoot corresponding to a damping factor ζ of 0.45. Although the overshoot is somewhat larger than considered ideal (10% to 15%) the system appears to be stable and anticipates a small amout of compensation may be necessary.

4.6.4   Second Simulation – Inclusion of Shaft Dynamics

Figure 4.52 shows the system with the shaft dynamics added per Equation 4.64.

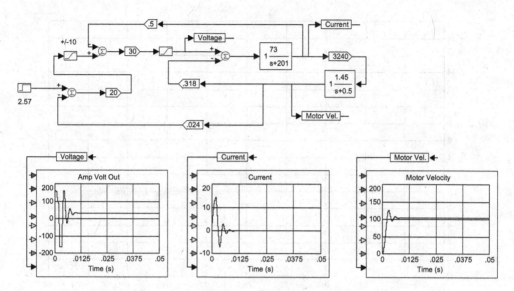

Figure 4.51   Initial simulation; motor and load inertia as single value

The system now shows extreme instability. Figure 4.53 shows two dominant frequencies: the "normal" system natural frequency plus the oscillation due to shaft compliance. The higher frequency, as measured on Figure 4.52, is 2800 Hz and as calculated per $f = \frac{1}{2\pi}\sqrt{\frac{K(J_m + J_l)}{J_m J_l}} =$ 2700 Hz are in close agreement and verify that the formula is a good rule-of-thumb to use to at least determine the possible cause of a problem. This system needs compensation to become stable.

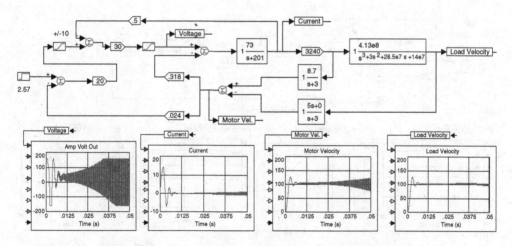

Figure 4.52   Second simulation; shaft dynamics added, oscillation

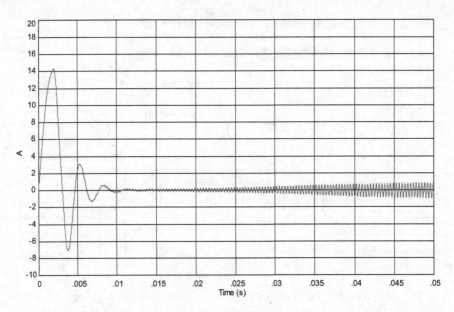

Figure 4.53   Second simulation showing two dominant frequencies

4.6.5   Third Simulation – Compensation

Figure 4.54 shows the system with a lead/lag compensation added, eliminating the oscillation and resulting in an acceptable velocity rise. However, placing the lead an octave below the troublesome frequency has actually over compensated the system.

Figure 4.55 has the lead increased to almost the same as the resonant frequency, 2400 compared to 2800, it still maintains stability with a more rapid rise and an overshoot of only 5%.

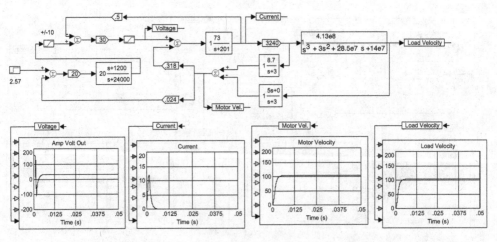

Figure 4.54   Third simulation; lead/lag compensated, over compensated

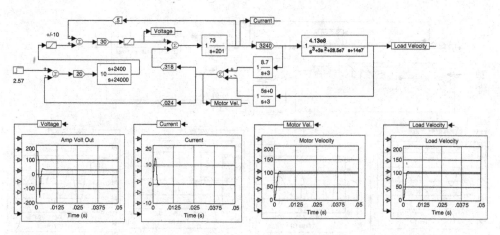

Figure 4.55 Third simulation; lead/lag compensated, optimal compensation (available in full color at www.wiley.com/go/moritz)

4.6.6 Coupling Simulation

Figure 4.56 shows the system modified to include a flexible coupling with a stiffness of $K = 1.6 \times 10^7$ g cm rad^{-1} compared to the solid shaft K of 2.73×10^7 g cm rad^{-1}.

Satisfactory performance still results, showing that for this system, if a coupling is needed, the one sampled here will be acceptable

A coupling with higher flexibility but with $K = 1.84 \times 10^5$ g cm rad^{-1}, two orders of magnitude below the previous sample, has unacceptable results, as shown in Figure 4.57.

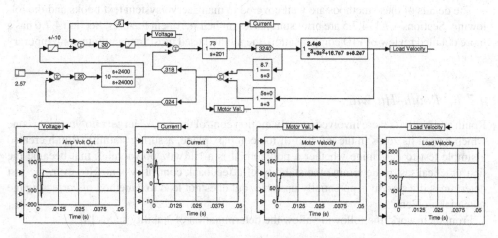

Figure 4.56 Third simulation with stiff flexible coupling (available in full color at www.wiley.com/go/moritz)

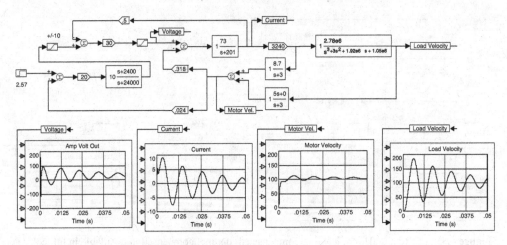

Figure 4.57   Third simulation with weak flexible coupling

4.7   Compensation

The area which has dominated the design of motion systems has been that of system compensation; the requirement to achieve required performance while maintaining system stability; eliminating oscillation and achieving acceptable response with minimum overshoot.

Basically, this has meant creating methods to efficiently examine the system characteristic equation, determine if potential instability exists and create required design modifications.

All of the early methods pre-date the digital computer and the high level software programs which now allow rapid simulation and evaluation of system design, replacing the laborious hand calculations required in the past. These early methods can also be created with the computer, only faster and more accurately, continuing to aid the designer in visualizing the system performance and evaluating various methods of optimization

The details of these methods are well covered in many servo/system text books and the following. Sections 4.7.1–4.7.5 are brief summaries of their pertinent features. Section 4.7.6 has a more detailed review of PID compensation, the method used most frequently in contemporary systems.

4.7.1   Routh–Hurwitz

Routh and Hurwitz were involved in early motion control theory. A Hurwitz polynomial is one which has all its zeros in the left half of the complex plane, assuring stability. Routh created a simple test to determine whether a polynomial is a Hurwitz polynomial, that became the first practical test to determine the stability of a feedback control system. However, the test determines if a system is potentially unstable, but does not lead to a method of correcting the potential problem.

From their work, the following allow the determination of stability status [8]:

- Write the characteristic equation in descending order of the powers of s.
- The degree of the characteristic equation will equal the number of roots.

- Complex roots always exist in conjugate pairs.
- For the real parts of all the roots to be negative (lie in the left half of the complex plane), all the coefficients must have the same sign (necessary but not sufficient).
- If all the coefficients do not have the same sign, or if not all terms are present in descending order of s, then one or more roots are positive and the system is unstable.

Proceeding from this overview, Routh developed a method of manipulating the coefficients to create an array, the Routh Array, (columns and rows) that resulted in a column of numbers. If all the numbers in this column are positive, the system is stable.

If one or more numbers in the column are negative, the system is unstable and the number of times the signs change is equal to the number of roots with positive real parts.

The main use of the Routh–Hurwitz evaluation is to determine the potential instability of a system and to aid in solving the characteristic equation, especially if it is of higher than third order [9] However, by using programs such as Poly Roots (see Section 2.1) it is now possible to determine the roots of any order of polynomial and accomplish both goals expeditiously.

4.7.2   Nyquist

The Nyquist diagram is a plot of the magnitude and phase on the complex plane of the open loop transfer function as the frequency varies from zero to infinity. It essentially examines the characteristic equation:

$$1 + \mathbf{GH} = 0$$

stated as:

$$\mathbf{GH} = -1$$

and shows the path the end point of the complex vector $\mathbf{GH}$ traces with respect to the $-1 + j0$ point on the real axis. If the trace encircles the -1 point, the system will be unstable. Where the path crosses a circle of radius 1 will determine the system phase margin and where it crosses the negative real axis will determine the gain margin.

Although the Nyquist diagram displays the same information as the Bode and Root Locus diagrams, it is not as forthcoming in suggesting how to add compensation to improve stability.

4.7.3   Bode

Bode diagrams, or gain/phase diagrams, are one of the most useful means of displaying the open loop frequency-dependent characteristics of a system, leading to a determination of what the closed loop dynamics will be. Bode created the concept of the gain and phase margin, which helps to determine if compensation is needed and at what frequencies (break points) leads, lags, and so on it should be placed by adding compensating networks in order to create stable closed loop performance. Once a system is computer simulated, its Bode plots can be displayed, printed and examined in much less time than was previously required by hand sketching.

4.7.4 Root Locus

The root locus diagram is a plot of the variation of the roots of the characteristic equation as one or more parameters of the system are varied. Usually, the gain is varied, at least initially, in order to determine if compensation is needed as the gain is increased from zero. If poles move from the left half to the right half of the complex plane, then instability exists and some parameter must be changed or zeros added (by compensating networks) to move the destabilizing poles back into the left half of the plane or cancel them. Drawing the root locus diagram requires following an extensive set of rules but again, once a system has been computer simulated, the root locus can be displayed quickly.

4.7.5 Phase Plane

The phase plane diagram is a plot of the system velocity versus position in response to a particular command (step, ramp, sine, etc.). Theoretically, analysis using the phase plane is limited to second order systems, since higher order systems would require plotting in higher dimensions. However, nothing prevents plotting system velocity versus position for any system, regardless of order. Generating the phase plane plot is very helpful in visualizing the dimensional requirements of a system, especially one in which the motion of the load is limited. The typical plot of load position versus time does not have the same visual impact as the same information plotted in the phase plane diagram. Also, the phase plane diagram is especially useful in showing the effect of nonlinearities, such as deadband and saturation.

4.7.6 PID

In 1922 Michael Minorsky described a "3 term" control scheme which we today call PID control. This compensation technique has become the most popular and widely used method for stabilizing motion systems, especially since the development of digital programable controllers make it easy to experiment with and adjust actual hardware.

The classic PID diagram is shown in Figure 4.58. This version of the PID filter is known as the non-interacting form since each term can be individually adjusted without affecting the other two terms.

The error is processed by the sum of the three terms of the PID according to

$$o(t) = e(t)\left[P + \frac{I}{S} + SD\right] \qquad (4.65)$$

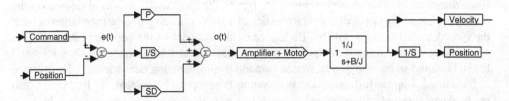

Figure 4.58 Block diagram of classic non-interacting PID compensation .

The P is a proportional term, usually referred to as the system gain. It essentially determines the stiffness of the system; the amount of torque or force experienced by the load as a result of a motion command.

The I term integrates the error and determines system accuracy. The output of an integrator will always increase if its input is non-zero. In a closed loop situation, it will drive the system error to zero.

The D term creates the derivative of the error, essentially providing the rate of change of the error, stabilizing the system, similar to the action of a tachometer.

A better understanding of the function of the PID block can be achieved as follows [10, 11]:

$$P + \frac{I}{S} + SD = \frac{D\left(S^2 + \frac{P}{D}S + \frac{I}{D}\right)}{S} \tag{4.66}$$

If $I = 0$,

$$\text{then the PID block} = D\left(S + \frac{P}{D}\right) \tag{4.67}$$

adding a zero to the open loop transfer function at $-\frac{P}{D}$ to aid in compensating one of the poles. If $D = 0$,

$$\text{then the PID block} = \frac{P}{S}\left(S + \frac{I}{P}\right) \tag{4.68}$$

adding a zero at $-\frac{I}{P}$ and a pole to the open loop transfer function, causing integration to reduce system error and compensate for one of the poles.

If all three terms are present, the numerator quadratic will create two zeros and the denominator S term will cause integration and error reduction.

Stating the PID terms in this form makes it possible to assign values to them in conjunction with performing design activity using Bode or root locus plots.

Various forms of PID architecture have evolved, depending on the design philosophy and experience of manufacturers. Three of the many variations are shown in Figure 4.59.

Some methods create a virtual velocity loop, then close a position loop around it, emulating the traditional two loop design.

A great deal of information has been published in industry trade journals and company white papers about the proper method to adjust, or tune, a PID-based system A non-inclusive survey located over 20 such articles published over a 10 year period, all of them describing various adjustment methods and all claiming to be the best and most efficient. Two of the best publications, based on practical "hands on" experience together with pertinent theory, are by Charles Raskin [12] and George Ellis [13].

Various formulas and procedures have been developed to aid in adjusting PID-based controllers, the most frequently used being the Ziegler/Nichols method, based on experience in the materials process industry. In addition, controllers now also include self-tuning algorithms which exercise the system and set the PID parameters with little or no manual intervention.

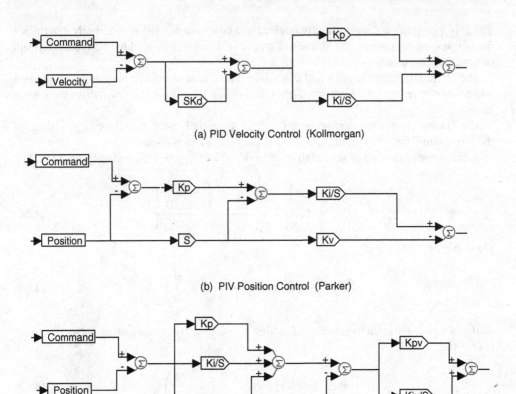

(a) PID Velocity Control (Kollmorgan)

(b) PIV Position Control (Parker)

(c) PID Cascade (PMD)

Figure 4.59   Three forms of PID compensation

The Ziegler/Nichols procedure consists of the following steps:

- Slowly increase system gain until the load starts to oscillate.
- Record the gain as P_{test} and the frequency in Hertz as f_{test}.
- Set $P = 0.6P_{test}$
- Set $I = 2f_{test}P$
- Set $D = P/8f_{test}$.

Figure 4.60 shows the system of Figure 4.11 re-configured for PID compensation.

The P was increased until output oscillations were observed, as shown, and the frequency measured as 5 Hz with $P_{test} = 2$.

$$\text{Therefore:} \quad P = 1.2$$
$$I = 12$$
$$D = 0.03$$

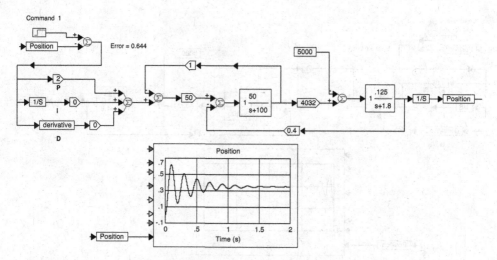

Figure 4.60   Figure 4.11 with PID added (available in full color at www.wiley.com/go/moritz)

The step response is shown in Figure 4.61

The system has a rise time of 0.2 s, an overshoot of 16%, a settling time of 0.5 s and the error has been reduced by two orders of magnitude below the uncompensated system.

A problem with this procedure is that it is subjective with regard to what constitutes the start of oscillation. The procedure was repeated resulting in a frequency of 10 Hz with $P_{test} = 8$.

$$\text{Therefore:} \quad P = 4.8$$
$$I = 96$$
$$D = 0.06$$

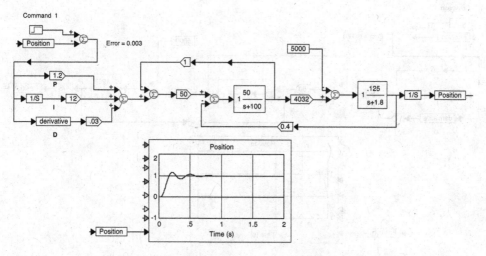

Figure 4.61   Figure 4.60 with initial PID adjustment

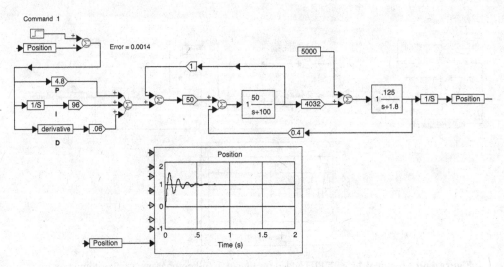

Figure 4.62   Figure 4.60 with increased PID values

The step response is shown in Figure 4.62.

The system now has a rise time of 0.05 s, an overshoot of 50%, a settling time of 0.5 s and an error of half the value of the first test.

Although this procedure does result in stable, improved performance the question arises as to whether further adjustment will produce even better performance. To test this, additional adjustments were made by increasing P and I while adjusting D to maintain stability. The final result is shown in Figure 4.63.

All results have improved, with a rise time of 0.015 s, overshoot of 20% and settling time of 0.1 s. The error has decreased another order of magnitude.

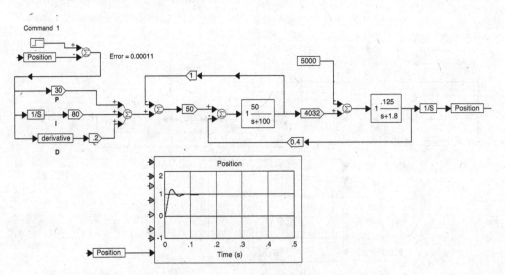

Figure 4.63   Figure 4.60 with final PID adjustment

4.7.7 Notch Filter

Along with PID capability, control software usually includes the ability to add notch filtering to the system, which is extremely useful to eliminate oscillations due to shaft or coupling compliance. The frequency of such oscillations is difficult to identify exactly during the design phase although an estimate is possible using values of the various inertias, shaft dimensions and coupling specifications. The actual frequency is typically determined during hardware testing Figures 4.52–4.55 show a system exhibiting a shaft compliance created oscillation and the use of a lead/lag filter to eliminate it.

Figure 4.64 shows the same system with a notch filter being used to eliminate the oscillation. One form of notch filter has the transfer function of:

$$\frac{S^2 + \omega^2}{S^2 + 2\zeta\omega + \omega^2} \tag{4.69}$$

where ω is the frequency to be suppressed and ζ is a factor defining the "sharpness' of the notch.

In Section 4.6, the compliance induced oscillation was calculated and tested to be 2700 Hz. With a chosen value of 0.2 for ζ, the notch filter has a transfer function of:

$$\frac{S^2 + 2.9e8}{S^2 + 6800S + 2.9e8} \tag{4.70}$$

The problem with the notch filter is that it works best for situations with fixed inertias. If the load inertia changes in normal system operation, then the oscillation will change also, making the notch filter ineffective. This can be somewhat alleviated by broadening the notch (using a larger value for ζ) but this then reduces the attenuation. If the load inertia change

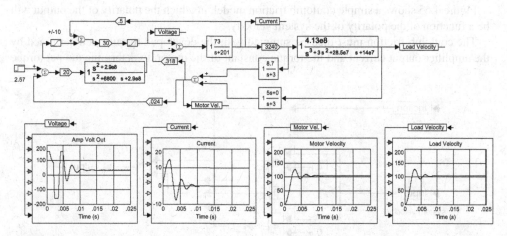

Figure 4.64 Figure 4.55 with notch filter replacing lead/lag compensation (available in full color at www.wiley.com/go/moritz)

can be anticipated, software modifications to the filter values would be possible. For a large variation in inertias the use of the lead/lag filter or other form of low pass filter would be most effective.

4.8 Nonlinear Effects

The ability to use simulation programs to model nonlinear characteristics has greatly advanced the system design process.

In the past, manual methods (describing function and phase plane) were developed to aid in determining the effect of nonlinearities, but were typically limited to one nonlinearity and to second order systems, plus they required the use of somewhat laborious calculations.

With computer simulation, multiple nonlinearities can be examined and their effect on overall performance can be observed while changing their characteristics while repeatedly running the program.

Of great interest during the design process is to run the simulation initially as a completely linear system and then add the various nonlinearities to determine their effect on system performance. For example, a linear model might show that the voltage output of the servo amplifier peaks at 300 V during acceleration. However, knowing that the system will be line operated at 120 V AC, resulting in a 160 V DC bus, a voltage limit block can be added to the simulation of the amplifier to determine the effect of such a limitation on performance. This might then show that the voltage constant of the motor has to be reduced and/or the peak current requirement from the amplifier has to be increased.

4.8.1 Coulomb Friction

Coulomb friction, typically referred to as just friction, is the torque or force that opposes the applied force or torque and is usually assumed to be independent of velocity. It opposes acceleration and aids deceleration.

Figure 4.65 shows a simple coulomb friction model, in which the polarity of the output will be a function of the polarity of the system velocity.

The coulomb friction must then be summed with the developed torque, as determined by the amplifier output current and the torque constant of the motor, to determine the net torque

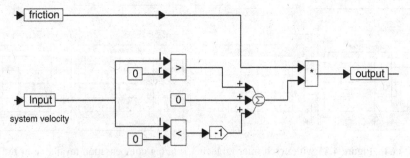

Figure 4.65 Coulomb friction model

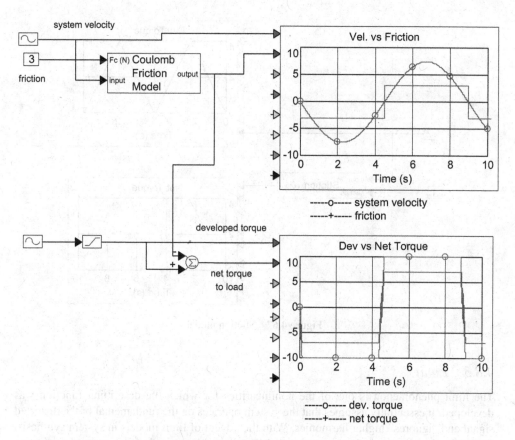

Figure 4.66 Coulomb friction and net torque

applied to the load, as shown in Figure 4.66. In the absence of any load torque, it is the coulomb friction and the viscous friction (*B*) that will determine the steady state current in the system.

4.8.2 Stiction

Unlike coulomb friction, which is always present, stiction (static friction) is only present until motion commences. It is larger than coulomb friction. The applied torque or force must "build up" to the stiction value and "break through" it, at which point total friction drops to the coulomb values. Figure 4.67 shows a friction block with a ramp of developed torque applied to it. The output remains at zero until the developed torque equals the stiction value, at which time the output then follows the applied torque.

A problem with this and other stiction models is that they assume that stiction appears each time the input passes through zero. Actually, in an oscillatory system, stiction does not necessarily have time to "build up" as zero velocity is crossed but will require some pause at zero velocity to make itself felt. In this model, based on empirical data, the stiction term could be held at zero for some time after zero velocity is reached.

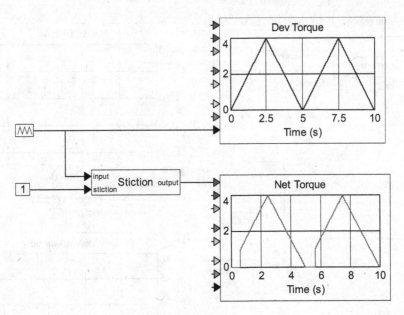

Figure 4.67   Stiction model

4.8.3   Limit

The limit phenomenon is one of the nonlinearities for which the describing function was developed. It essentially assumes that the system operates on the fundamental of the distorted signal and "ignores" higher harmonics. With the advent of limit models in system synthesis, this approximation is no longer applicable. The limit block produces the complete system response to the limit.

A limit block is very useful in determining the maximum output of a servo amplifier, as determined by the system bus voltage. Once the limit is reached, amplifier output will no longer be proportional to the amplifier input, effectively opening the loop until conditions cause the voltage to fall below the limit value. Figure 4.68a shows conditions for an amplifier with a gain of 20 and an input of 15 V with and without a limit.

An additional use for the limit block, also for amplifier simulation, is to limit the output current of a current drive amplifier to its rated value. Since the gain of such an amplifier is given by A V^{-1}, limiting the input voltage will directly result in limiting the output current. Figure 4.68b shows an amplifier with a gain of 20 A V^{-1} with an input signal of 2 V, with and without current limiting.

4.8.4   Deadband

Deadband is a zone in an assembly within which an input excursion does not create a corresponding output excursion. Once the input reaches the limit of the dead zone the output then follows the input. Figure 4.69 shows a deadband block with two levels of input and the same deadband setting, illustrating the effect. Note in this simulation that once the input exceeds the deadband, the output then follows. This would only be the case if the output could

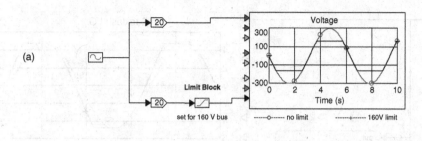

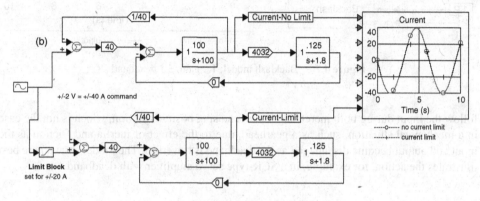

Figure 4.68   Limit model

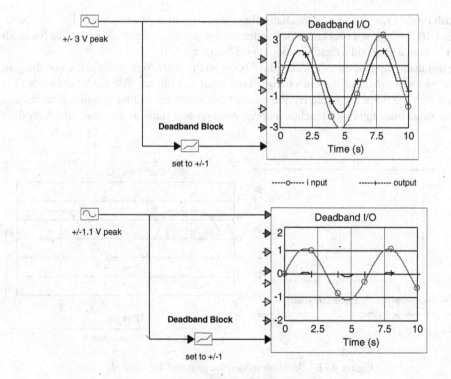

Figure 4.69   Deadband model

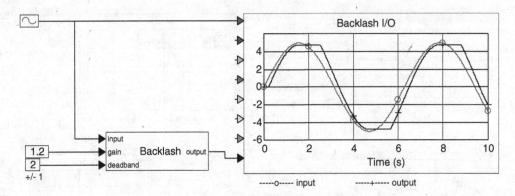

Figure 4.70   Backlash model; 1.2 gain, ±1 deadband

follow the input during both increasing and decreasing position motion, which is not the case in a mechanical situation, such as a gearhead, due to the effect of inertia and friction as the input and output become disengaged as the deadband is traversed. The model shown here best illustrates the action, for example, in a SCR-type servo amplifier with deadband.

4.8.5   Backlash

Backlash models the action in a mechanical system, such as a gearhead, that has deadband. Figures 4.70–4.72 show a backlash block under various input conditions. Note that the backlash block has both a gain and a deadband input (see Section 4.8.4).

The deadband input allows the "free play" between the gears to be modeled, while the gain input allows adjustment of the ratio between the input and output. The model shown here is one in which the output has relatively high friction, where the output position remains at a constant value once the input reaches its peak position and starts to decrease, at which time

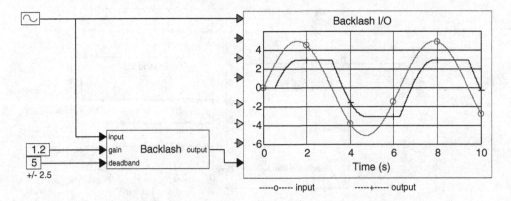

Figure 4.71   Backlash model; 1.2 gain, ±2.5 deadband

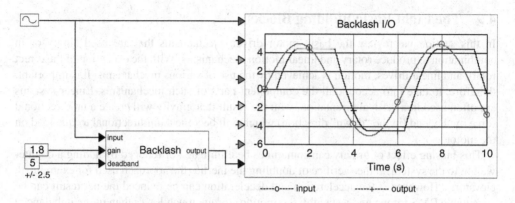

Figure 4.72   Backlash model; 1.8 gain, ±2.5 deadband

the input loses contact with the output. Once the input traverses the deadband, contact is again made with the output and the cycle repeats.

If the output has relatively high inertia, then the output would continue to move in a positive direction as the input backs away and contact would again be made once the deadband gap is closed by the relative motion of the two parts [14].

4.8.6   Hysteresis

The backlash described in Section 4.8.5 has hysteresis, in that as the input first increases and then decreases, the output remains at its maximum value until the input increases in the opposite direction by an amount to again make contact with the output member and move it to a new maximum.

Figure 4.73 shows the extreme case of hysteresis in which the output moves between two extremes (plus and minus 1) as the input moves both above and below the high and low hysteresis limits. Note that the limits are independently settable, allowing the simulation of unsymmetrical hysteresis.

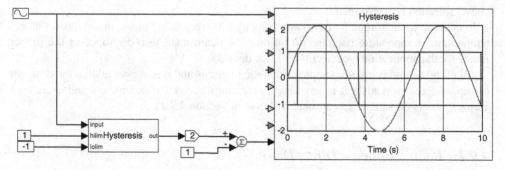

Figure 4.73   Hysteresis model

4.9   The Eight Basic Building Blocks

In this section we review the basic motor driven mechanisms that are used singly or in combination to produce rotary and linear motion mechanisms. With the exception of the direct rotary and linear drives, motion is achieved by the use of various mechanisms. It is important, therefore, to take into account all the component parts of such mechanisms. Linear systems are often used in a vertical orientation, with the result that gravity will create a unidirectional force on the load (in the "down" direction) which will become a unidirectional torque load on the motor.

This biasing effect of gravity can sometimes be eliminated or reduced by adding a counter weight to the system, at the sacrifice of doubling the inertia (the approach used for example, in elevators). However, if the acceleration and deceleration can be reduced the net result can be acceptable RMS torque and a standby (no motion) torque much lower than in the unbalanced configuration. In addition, the counterbalance approach reduces the size of the fail-safe or holding brake should one be required.

In each case, equations are provided showing how to calculate the acceleration, deceleration, run and standby torques and corresponding velocities required from the motor in response to system velocity profile specifications. The torques are based on the assumption of a trapezoidal velocity profile, which results in constant current for each of the four parts of the profile. Once these torques are determined, a preliminary motor selection and then the four currents and the RMS current can be calculated. For all the systems, the RMS torque will be:

$$T_{RMS} = \sqrt{\frac{T_{ACC}^2 t_{ACC} + T_{RUN}^2 t_{RUN} + T_{DEC}^2 t_{DEC} + T_{SS}^2 t_{SS}}{t_{ACC} + t_{RUN} + t_{DEC} + t_{SS}}}$$

If other profiles (hyperbolic, sine, S, etc.) are used, the acceleration and deceleration currents will not be constant but functions of time, which fact must be included in the RMS calculation.

Note, however, that one part of the total inertia seen by the motor is the motor inertia itself, which initially is not known.

A way to begin is to use all the known inertias to calculate an initial acceleration torque, which will typically be the largest of the four torques. Use this torque to review motor catalogs and/or have a discussion with motor supplier application engineers to make an initial motor selection. Repeat the calculations to determine if the motor selected was too "small", too "large" or satisfactory. The calculations may have to be repeated a number of times until a "best" selection is determined.

In addition to the torque (current) and velocity (voltage) calculations, dissipation, ambient temperature, temperature rise, and so on must be determined in order to select the proper motor for the application (see Section 3.1. for details).

All of the formulas involve simple arithmetic relations and it is suggested that the designer set up an Excel spreadsheet to expedite calculating the system parameters and evaluating various motors (A sample spreadsheet is shown in Section 4.9.1).

4.9.1   Rotary Motion – Direct Drive

See Figure 4.74.

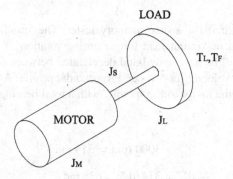

Figure 4.74 Rotary motion; direct coupled

System Data:

$$J_M = \text{motor inertia (g cm s}^2)$$
$$J_L = \text{load inertia (g cm s}^2)$$
$$J_S = \text{shaft inertia (g cm s}^2)$$
$$T_L = \text{load torque (g cm)}$$
$$T_F = \text{friction torque (g cm)}$$

Motion:

$$\text{Position: } \theta_M = \theta_L \text{ (rad)}$$
$$\text{Velocity: } \theta'_M = \theta'_L \text{ (rad s}^{-1})$$
$$\text{Acc/Dec: } \theta''_M = \theta''_L \text{ (rad s}^{-2})$$

At Motor:

$$J_T \text{ (total inertia)} = J_M + J_L + J_S \tag{4.71}$$
$$T_{ACC} = +J_T\theta''_M + T_F \pm T_L \tag{4.72}$$
$$T_{DEC} = -J_T\theta''_M + T_F \pm T_L \tag{4.73}$$
$$T_{RUN} = +T_F \pm T_L \tag{4.74}$$
$$T_{SS} = \pm(T_L - T_F) \quad \text{if} \quad T_L > T_F \tag{4.75}$$
$$= 0 \text{ if } T_L < T_F$$

The sign for T_L depends on whether it is bi-or unidirectional and whether it opposes or aids acceleration or deceleration. At zero velocity, if a load torque is present and it is greater than friction torque, then the motor must develop a counter torque to maintain zero velocity.

Example

A spindle drive is required for a disc memory tester. The spindle inertia (a cylinder) is 5.4 g cm s^2. There is a maximum load torque during rotation of 720 g cm opposing the rotation. The spindle will be accelerated and decelerated between 0 and 3000 rpm in 0.5 s, run for 5 s and rest at 0 velocity for 5 s, per a trapezoidal profile. A coupling/shaft assembly will be used to connect the motor to the spindle, with a total bearing torque of 144 g cm and inertia of 0.072 g cm s^2

$$\theta'_M = 3000 \text{ rpm} = 314 \text{ rad s}^{-1}$$

$$\theta''_M = 314/0.5 = 628 \text{ rad s}^{-2}$$

Since this is a fairly small system, in order to quickly get a sense of the accelerate torque involved, we will initially just use the known inertias.

$$\text{Initial inertia} = 5.4 + 0.072 = 5.5 \text{ g cm s}^2$$

$$\text{Initial } T_{ACC} = (5.5)(628) + 720 + 144 = 4318 \text{ g cm}$$

First motor choice: A NEMA size 23 brushless servo motor with the following specs.

$T_{CONT} = 5616 \text{ g cm} \quad T_{PEAK} = 16\,920 \text{ g cm}$

$N_{MAX} = 4000 \text{ rpm} \quad R = 11.9\ \Omega$

$K_T = 2880 \text{ g cm A}^{-1} \quad J_M = 0.0013 \text{ oz in s}^2$

$J_T = 0.094 + 5.4 + 0.072 = 5.6 \text{ g cm s}^2$

$T_{ACC} = (5.6)(628) + 720 + 144 = 4381 \text{ g cm}$ $I_{ACC} = 4381/2880 = 1.5 \text{ A}$

$T_{DEC} = -(5.6)(628) + 720 + 144 = -2653 \text{ g cm}$ $I_{DEC} = 2653/2880 = 0.92 \text{ A}$

$T_{RUN} = 720 + 144 = 864 \text{ g cm}$ $I_{RUN} = 864/2880 = 0.3 \text{ A}$

$$T_{RMS} = \sqrt{\frac{4381^2(0.5) + 864^2\,(5) + 2653^2(0.5)}{0.5 + 5 + 0.5 + 5}} = 1237 \text{ g cm} \quad I_{RMS} = 1237/2880 = 0.43 \text{ A}$$

Although this motor can easily do this application, it appears to be quite oversized and it would be advantageous to see if a smaller, less expensive motor is available.

Second motor choice: A NEMA size 16 brushless servo motor with the following specs.

$$T_{CONT} = 2664 \text{ g cm} \quad T_{PEAK} = 7992 \text{ g cm}$$

$$N_{MAX} = 5000 \text{ rpm} \quad R = 4.38\ \Omega$$

$$K_T = 1015 \text{ g cm A} \quad J_M = 0.022 \text{ g cm s}^2$$

Repeating the calculations results in:

$$J_T = 5.5 \text{ g cm s}^2$$
$$T_{ACC} = 4318 \text{ g cm} \qquad I_{ACC} = 4.3 \text{ A}$$
$$T_{DEC} = -2590 \text{ g cm} \qquad I_{DEC} = 2.6 \text{ A}$$
$$T_{RUN} = 864 \text{ g cm} \qquad I_{RUN} = 0.85 \text{ A}$$
$$T_{RMS} = 1221 \text{ g cm} \qquad I_{RMS} = 1.2 \text{ A}$$

This motor can also do the application, has an almost 2:1 safety margin for both the maximum and RMS currents, a 1.7:1 margin for the maximum speed and would be less expensive than the size 23 first chosen.

Table 4.3 outlines an Excel program for the above calculations and Table 4.4 shows the results of running this program for the first choice motor. It can easily be modified to develop a program for each of the remaining building blocks shown in this section.

Table 4.3 Excel program for direct drive calculations

	A	B
1		
2		**TABLE 4.3**
3		
4		**Excel Program for Example in 4.9.1**
5		
6	SYSTEM SPECS	
7		
8	Load Inertia	5.4
9	Shaft Inertia	0.072
10	Motor Velocity (max)	3000
11	Acc Time	0.5
12	Dec Time	0.5
13	Run Time	5
14	Stop Time	5
15	Acc	=+((B10/60)*2*3.1416)/B11
16	Dec	=+((B10/60)*2*3.1416)/B12
17	Friction Torque	144
18	Load Torque	720
19		
20	MOTOR SPECS	
21		
22	Torque (cont)	5616
23	Torque (peak)	16 920
24	Max Velocity	4000
25	Resistance	11.9
26	Torque Constant	2880
27	Inertia	0.094
28		
29	SYSTEM DATA	
30		
31	Total Inertia	=+B8+B9+B27
32	Acc Torque	=+B31*B15+B17+B18
33	Dec Torque	=-B31*B16+B17+B18
34	Run Torque	=+B17+B18
35	RMS Torque	=+((((B32)^2*B11)+((B34)^2*B13)+((B33)^2*B12))/(B11+B12+B13+B14))$^{0.5}$
36	Acc Current	=+B32/B26
37	Dec Current	=+B33/B26
38	Run Current	=+B34/B26
39	RMS Current	=+B35/B26

Table 4.4 Results using Excel program in Table 4.3 for first motor selection Example in 4.9.1

	A	B
1		
2	**TABLE 4.4**	
3		
4	**Results for First Motor Selection**	
5		
6	SYSTEM SPECS	
7		
8	Load Inertia	5.4
9	Shaft Inertia	0.072
10	Motor Velocity (max)	3000
11	Acc Time	0.5
12	Dec Time	0.5
13	Run Time	5
14	Stop Time	5
15	Acc	628.32
16	Dec	628.32
17	Friction Torque	144
18	Load Torque	720
19		
20	MOTOR SPECS	
21		
22	Torque (cont)	5616
23	Torque (peak)	16 920
24	Max Velocity	4000
25	Resistance	11.9
26	Torque Constant	2880
27	Inertia	0.094
28		
29	SYSTEM DATA	
30		
31	Total Inertia	5.566
32	Acc Torque	4361.23
33	Dec Torque	−2633.23
34	Run Torque	864
35	RMS Torque	1232.50
36	Acc Current	1.51
37	Dec Current	−0.91
38	Run Current	0.30
39	RMS Current	0.43

4.9.2 Rotary Motion – Gearhead Drive

N is the gear ratio of the gearhead. For most applications, this ratio is greater than 1 in order to create a condition in which the motor is operating at as high an efficiency as possible.

For gearheads in which N is 10 or greater, the load inertia reflected back to the motor will often be relatively small, in which case a good initial approximation for the total inertia seen by the motor is $J_M + J_{GH}$. See Figure 4.75.

System Data:

$$J_M = \text{motor inertia (g cm s}^2)$$
$$J_L = \text{load inertia (g cm s}^2)$$
$$J_{GH} = \text{gearhead inertia (g cm s}^2)$$
$$J_S = \text{shaft inertia (g cm s}^2)$$
$$T_L = \text{load torque (g cm)}$$
$$T_F = \text{friction torque (g cm)(seen at gearhead input)}$$

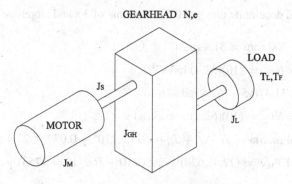

Figure 4.75 Rotary motion; gearhead coupled

Gearhead Efficiency (e): Spur or Bevel = 0.90; Worm = 0.65; Planetary = 0.95
  Motion:

$$\text{Position: } \theta_M = N\theta_L \text{ (rad)}$$
$$\text{Velocity: } \theta_M' = N\theta_L' \text{ (rad s}^{-1})$$
$$\text{Acc/Dec: } \theta_M'' = N\theta_L'' \text{ (rad s}^{-2})$$

At Motor:

$$J_T \text{ (total inertia)} = J_M + J_L/N^2 + J_{GH} \tag{4.76}$$

(theoretical, assuming the gearhead efficiency is 100%)
  In order to account for the actual gearhead efficiency, the torques are calculated as follows:

$$T_{ACC} = (J_M + J_{GH})\theta_M'' + (J_L/N^2 e)\theta_M'' \pm T_L/Ne + T_F \tag{4.77}$$
$$T_{DEC} = -(J_M + J_{GH})\theta_M'' - (J_L/N^2 e)\theta_M'' \pm T_L/Ne + T_F \tag{4.78}$$
$$T_{RUN} = +T_F \pm T_L/Ne \tag{4.79}$$
$$T_{SS} = \pm(T_L/Ne - T_F) \text{ if } T_L/Ne > T_F \tag{4.80}$$
$$= 0 \text{ if } T_L/Ne < T_F$$

The sign for T_L depends on whether it is bi-or unidirectional and whether it opposes or aids acceleration or deceleration. At zero velocity, if a load torque is present and it is greater than friction torque, the motor must develop a counter torque to maintain zero velocity.

Example
A system with a load inertia of 2736 g cm s^2 must be accelerated and decelerated between 0 and 300 rpm in 0.5 s. A spur gearhead with a ratio of 10:1 has been chosen so as not to exceed a maximum gearhead input speed of 3200 rpm. The gearhead has an input inertia of 0.072 g cm s^2. There is 2880 g cm of system friction torque at the load and a bidirectional load torque of 360 g cm opposing the direction of rotation. A trapezoidal profile will be used

with accelerate and decelerate times of 0.5 s, run time of 3 s and stop time of 5 s.

$$\theta'_L = 300 \text{ rpm} = 31.4 \text{ rad s}^{-1}$$

$$\theta'_M = N\theta'_L = (10)(31.4) \text{ rad s}^{-1}$$

$$\theta''_L = 31.4/0.5 = 62.8 \text{ rad s}^{-2}$$

$$\theta''_M = N\theta''_L = (10)(62.8) = 628 \text{ rad s}^{-2}$$

Initial inertia $= J_L/N^2 + J_{GH} = 2736/10^2 + 0.072 = 27.4 \text{ g cm s}^2$

Initial $T_{ACC} = (27.4)(628) + 2880/10 + 360/10 = 17531 \text{ g cm}$

Motor Choice: A NEMA size 34 brushless servo motor with the following specs:

$$T_{CONT} = 7920 \text{ g cm} \qquad T_{PEAK} = 26712 \text{ g cm}$$

$$N_{MAX} = 5400 \text{ rpm} \qquad R = 4.8 \, \Omega$$

$$K_T = 3526 \text{ g cm A}^{-1} \quad J_M = 0.15 \text{ g cm s}^2$$

Gearhead Choice: a spur gearhead with the following specs:

$$T_{ACC} \text{ (output)} = 345\,600 \text{ g cm} \qquad T_{NOM} \text{ (output)} = 253\,440 \text{ g cm}$$

$$N_{IN} = 3200 \text{ rpm nom; } 6000 \text{ rpm max}$$

$$J_{GH} = 0.072 \text{ g cm s}^2 \qquad\qquad \text{eff.} = 90\%$$

$$T_{ACC} = (0.15 + 0.072)(628) + (2736/(10^2 \times 0.9))(628) + 2880/(10 \times 0.9) + 360/(10 \times 0.9)$$

$$= 19\,590 \text{ g cm}$$

$$T_{DEC} = -(0.15 + 0.072)(628) - \left(2736/(10^2 \times 0.9)\right)(628) + 2880/(10 \times 0.9) + 360/(10 \times 0.9)$$

$$= -18\,871 \text{ g cm}$$

$$T_{RUN} = (2880 + 360)/(10 \times 0.9) = 369 \text{ g cm}$$

Since $T_F > T_L$

$$T_{SS} = 0$$

$$T_{RMS} = \sqrt{\frac{19\,590^2(0.5) + 360^2(3) + 18\,871^2(0.5)}{0.5 + 3 + 0.5 + 5}} = 6415 \text{ gm cm}$$

$$I_{ACC} = 19\,590/3528 = 5.6 \text{ A} \quad I_{DEC} = 18\,871/3528 = 5.3 \text{ A}$$

$$I_{RUN} = 360/3528 = 0.1 \text{ A} \qquad I_{RMS} = 6415/3528 = 1.8 \text{ A}$$

The torques and speed are within the specs of both the motor and gearhead, although the RMS torque only has an 18% safety margin with respect to the motor continuous torque rating and possibly a somewhat larger motor should be considered.

T_{ACC} at load $= (2736)(62.8) = 171\,820 \text{ g cm}$; a 2 : 1 safety margin for the gearhead

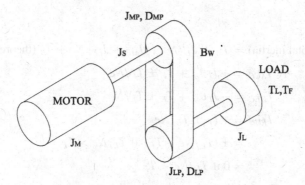

Figure 4.76   Rotary motion; belt and pulley coupled

4.9.3   Rotary Motion – Belt and Pulley Drive

The belt and pulley drive is similar to the gearhead drive in that there is a ratio between the motor and the load. In this case, N is the ratio of the load pulley diameter (D_{LP}) to the motor pulley diameter (D_{MP}).

$$N = D_{LP}/D_{MP}$$

In addition, the weight of the belt must be reflected to the motor as an inertia, especially if the belt is long or of a high mass material such as a chain link belt. See Figure 4.76.

System Data:

J_M = motor inertia (g cm s^2)

J_L = load inertia (g cm s^2)

J_{MP} = motor pulley inertia (g cm s^2)

J_{LP} = load pulley inertia (g cm s^2)

J_S = shaft inertia (g cm s^2)

T_L = load torque (g cm)

T_F = friction torque (g cm) (seen at motor pulley)

B_W = belt weight (g cm)

Belt weight reflected to the motor as an equivalent inertia:

$$J_B = \left(\frac{B_W}{980.6}\right)\left(\frac{D_{MP}}{2}\right)^2$$

Motion:

Position: $\theta_M = N\theta_L$ (rad)

Velocity: $\theta'_M = N\theta'_L$ (rad s^{-1})

Acc/Dec: $\theta''_M = N\theta''_L$ (rad s^{-2})

At Motor:

$$J_T \text{(total inertia)} = J_M + J_L/N^2 + J_{MP} + J_{LP}/N^2 + J_B \text{ (theoretical)} \tag{4.81}$$

$$T_{ACC} = +J_T\theta_M'' + T_F \pm T_L/Ne \tag{4.82}$$

$$T_{DEC} = -J_T\theta_M'' + T_F \pm T_L/Ne \tag{4.83}$$

$$T_{RUN} = +T_F \pm T_L/Ne \tag{4.84}$$

$$T_{SS} = \pm (T_L/Ne - T_F) \text{ if } T_L/Ne > T_F \tag{4.85}$$

$$= 0 \text{ if } T_L/Ne < T_F$$

The sign for T_L depends on whether it is bi- or unidirectional and whether it opposes or aids acceleration or deceleration. At zero velocity, if a load torque is present and it is greater than the friction torque, then the motor must develop a counter torque to maintain zero velocity.

Pulley efficiencies (e): timing belt $= 0.97$; chain and sprocket $= 0.96$

Example
A system with a motor pulley diameter of 1.27 cm and a width of 2.54 cm plus a load pulley diameter of 5.08 cm and a width of 2.54 cm is to accelerate an inertial load of 144 g cm s^2 from 0 to 600 rpm in 0.3 s, run for 5 s, decelerate to 0 rpm in 0.3 s and remain at 0 rpm for 2 s. There is 720 g cm of system friction torque at the motor shaft and a load torque of 1440 g cm in the direction of rotation. A timing belt with a weight of 227 g will be used.

$$J_{MP} = \frac{\rho\pi W R^4}{2g} = 0.0051 \text{ g cm s}^2$$

$$J_{LP} = 0.018 \text{ g cm s}^2$$

$$\theta_L' = 600 \text{ rpm} = 62.8 \text{ rad s}^{-1}$$

$$\theta_M' = N\theta_L' = (2/0.5)(62.8) = 251.3 \text{ rad s}^{-1}$$

$$\theta_M'' = 251.3/0.3 = 837.8 \text{ rad s}^{-2}$$

$$J_B = (227/980.6)(1.27/2)^2 = 0.093 \text{ g cm s}^2$$

Initial inertia $= 0.0051 + 0.093 + (1.3 + 144)/4^2 = 9.2 \text{ g cm s}^2$

Initial $T_{ACC} = (837.8)(9.2) + 720 - 1440/4 = 8068 \text{ g cm}$

Motor Choice: A NEMA 23 brushless servo motor with the following specs.

$T_{CONT} = 3816 \text{ g cm}$ $T_{PEAK} = 11\,520 \text{ g cm}$

$N_{MAX} = 5000 \text{ rpm}$ $R = 1.22 \, \Omega$

$K_T = 635 \text{ g cm A}^{-1}$ $J_M = 0.053 \text{ g cm s}^2$

$T_{ACC} = +(837.8)\left(0.053 + 0.0051 \times 10^{-4} + 0.093\right) + (837.8)\left((1.3 + 144)/(4^2 \times 0.97)\right)$

$\qquad + 720 - 1440/(4 \times 0.97) = +8319 \text{ g cm}$

$$T_{DEC} = -(837.8)(0.053 + 0.0051 + 0.093) - (837.8)\left((1.3 + 144)/(4^2 \times 0.97)\right)$$
$$+ 720 - 1440/(4 \times 0.97) = -7621 \text{ g cm}$$

$$T_{RUN} = +720 - 1440/(4 \times 0.97) = 349 \text{ g cm}$$

$$T_{SS} = 0$$

$$T_{RMS} = \sqrt{\frac{8319^2(0.3) + 349^2(5) + 7621^2(0.3)}{0.3 + 5 + 0.3 + 2}} = 2259 \text{ g cm}$$

$$I_{ACC} = 8319/635 = 13.1 \text{ A} \quad I_{DEC} = 7621/635 = 12 \text{ A}$$

$$I_{RUN} = 349/635 = 0.55 \text{ A} \quad I_{RMS} = 2259/635 = 3.6 \text{ A}$$

4.9.4 Linear Motion – Leadscrew/Ballscrew Drive

In evaluating a screw drive all of the following must be considered:

Screw Inertia (J_S): This can be determined using the inertia formula for a cylinder. Quite often, for a high pitch screw made of steel, the screw inertia is much larger than the reflected load inertia and the reflected load inertia can be ignored for initial calculations.

Friction Force (F_F): This is the opposing force created by the friction between the load and the load bearing surface. Do not confuse the coefficient of friction (μ) with the screw efficiency (e).

Nut Preload (T_P): To eliminate backlash, the drive nut, through which the screw rotates, is sometimes preloaded. This preload creates an additional torque load on the motor.

Note: When performing screw calculations do not confuse the screw pitch (P_S) which has the units of rev/cm with the screw lead (L_S) which has the units of cm/rev. (See Figure 4.77).

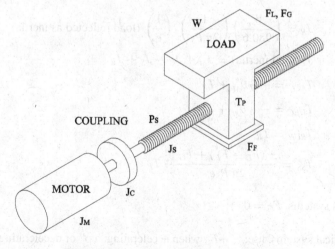

Figure 4.77 Linear motion; ballscrew drive

System Data:

$$J_M = \text{motor inertia (g cm s}^2)$$
$$J_C = \text{coupling inertia (g cm s}^2)$$
$$W = \text{load weight (g)}$$
$$T_P = \text{preload torque (g cm)}$$
$$F_L = \text{load force (g)}$$
$$F_G = \text{gravity force} = W \text{ (g)}$$
$$F_F = \text{friction force} = \mu W \text{ (g)}$$
$$J_S = \text{screw inertia (g cm s}^2)$$
$$T_B = \text{bearing torque (g cm)}$$
$$P_S = \text{screw pitch (rev/cm)}$$
$$e = \text{screw efficiency}$$
$$\mu = \text{friction coefficient}$$

Motion:

Position: $\theta_M = (2\pi P_S)(S)$ (rad)

Velocity: $\theta'_M = (2\pi P_S)(S')$ (rad s^{-1})

Acc/Dec: $\theta''_M = (2\pi P_S)(S'')$ (rad s^{-2})

At Motor:

$$T_R = T_P + T_B + \frac{\pm F_L + F_F \pm F_G}{2\pi P_S e} \text{(reflected torque)}$$

$$J_R = \left(\frac{W}{980.6}\right)\left(\frac{1}{2\pi P_S}\right)^2\left(\frac{1}{e}\right) \text{ (load reflected as inertia)}$$

$$J_T\text{(total inertia)} = J_M + J_C + J_S + J_R$$

$$T_{ACC} = +J_T\theta''_M + T_R$$

$$T_{DEC} = -J_T\theta''_M + T_R$$

$$T_{RUN} = T_R$$

$$T_{SS} = \frac{\pm(F_L - F_F) + F_G}{2\pi P_S e} - T_P$$

For horizontal systems, $F_G = 0$

For vertical systems, use: $+F_G$ when accelerating "up" or decelerating "down"

                            $-F_G$ when accelerating "down" or accelerating "up"

Coefficient of friction (μ)		Efficiency (e)	
Steel on steel	0.580	Ball nut	0.90
Steel on steel (lub)	0.150	Ball nut (preloaded)	0.80
Aluminum on steel	0.450	Acme w/metal nut	0.40
Brass on steel	0.350	Acme w/plastic nut	0.65
Dove-tail slide	0.200		
Ball bushing	0.001		
Linear bearing	0.001		
Teflon on steel	0.400		

Example

A mechanism to be used in an injection molding machine is required to move a 45 360 g load at a speed of 12.7 cm s^{-1} for a 107 cm travel. A linear bearing slide assembly will be used with a coefficient of friction of 0.001. A 1.97 pitch lead screw, 1.27 cm diameter by 122 cm long with preloaded ball nut will drive the load. The preload will be 1440 g cm.

Acceleration and deceleration times are to be 0.2 s. Zero velocity times between motions will be 5 s minimum. The assembly will have an estimated bearing friction torque of 360 g cm. A coupling with an inertia of 0.029 g cm s^2 has been selected.

$$J_S = \frac{\rho \pi L R^4}{2g} = 0.252 \text{ g cm s}^2$$

$$J_R = (45\,360/980.6)\,(1/(2\pi \times 1.97))^2\,(1/0.8) = 0.377 \text{ g cm s}^2$$

Note how the screw inertia and the reflected load inertia are the same order of magnitude.

$\theta'_M = (2\pi)(12.7)(1.97) = 157 \text{ rad s}^{-1} = 1500 \text{ rpm}$

$\theta''_M = 157/0.2 = 785 \text{ rad s}^{-2}$

$F_F = (0.001)(45\,360)(16) = 45.4 \text{ g}$

Initial inertia $= 0.029 + 0.252 + 0.377 = 0.658 \text{ g cm s}^2$

Initial $T_{ACC} = (0.658)(785) + 1440 + 360 + (45.4)\,(1/\,(2\pi \times 1.97 \times 0.8)) = 2321 \text{ g cm}$

Motor Choice: A NEMA 23 brushless servo motor with the following specs:

$T_{CONT} = 3888 \text{ g cm} \quad T_{PEAK} = 11\,592 \text{ g cm}$

$N_{MAX} = 5000 \text{ rpm} \quad R = 4.57 \ \Omega$

$K_T = 1236 \text{ gm cm/amp} \quad J_M = 0.053 \text{ gm cm sec}^2$

$T_{ACC} = +(0.658 + 0.053)\,(785) + 1440 + 360 + (45.4)\,(1/(2\pi \times 1.97 \times 0.8)) = +2363 \text{ g cm}$

$T_{DEC} = -(0.658 + 0.053)\,(785) + 1440 + 360 + (45.4)\,(1/(2\pi \times 1.97 \times 0.8)) = +1246 \text{ g cm}$

$T_{RUN} = 1805 \text{ g cm}$

$T_{SS} = 0$

In order to determine T_{RMS}, the run time must be calculated from the specifications that is, total travel is 107 cm, run velocity is 12.7 cm s^{-1} and acceleration and deceleration times are 0.2 s, therefore:

$$(0.5)(0.2)(12.7)(2) + (12.7)(t_{RUN}) = 107$$

$$t_{RUN} = 8.2 \text{ s}$$

$$T_{RMS} = \sqrt{\frac{2363^2(0.2) + 1805^2(8.2) + 1246^2(0.2)}{0.2 + 8.2 + 0.2}} = 1809 \text{ g cm}$$

$$I_{ACC} = 2363/1236 = 2 \text{ A} \quad I_{DEC} = 1246/1236 = 1.0 \text{ A}$$

$$I_{RUN} = 1805/1236 = 1.5 \text{ A} \quad I_{RMS} = 1809/1236 = 1.5 \text{ A}$$

4.9.5   Linear Motion – Belt and Pulley Drive

Similar to the belt and pulley drive for rotary load motion, the inertias of the pulleys and the weight of the belt must be considered in determining the total motor load (see Figure 4.78).
   System Data:

$$J_M = \text{motor inertia (g cm s}^2)$$
$$J_C = \text{coupling inertia (g cm s}^2)$$
$$W = \text{load weight (g)}$$
$$F_L = \text{load force (g)}$$
$$F_G = \text{gravity force} = W(g)$$
$$F_F = \text{friction force} = \mu W(g)$$
$$B_W = \text{belt weight(g)}$$
$$J_{PM} = \text{motor pulley inertia (g cm s}^2)$$

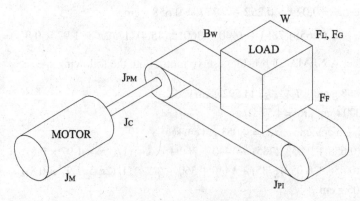

Figure 4.78   Linear motion; belt and pulley drive

$$J_{PI} = \text{idler pulley inertia (g cm s}^2)$$
$$e = \text{screw efficiency}$$
$$\mu = \text{friction coefficient}$$

Motion:

$$\text{Position: } \theta_M = \frac{S}{D_{PM}/2} \text{ (rad)}$$

$$\text{Velocity: } \theta'_M = \frac{S'}{D_{PM}/2} (\text{rad s}^{-1})$$

$$\text{Acc/Dec: } \theta''_M = \frac{S''}{D_{PM}/2} (\text{rad s}^{-2})$$

Belt weight reflected to motor as an equivalent inertia:

$$J_B = \left(\frac{B_W}{980.6}\right) \left(\frac{D_{PM}}{2}\right)^2 \left(\frac{1}{e}\right)$$

At Motor:

$$T_R = (\pm F_L + F_F \pm F_G) \left(\frac{D_{PM}}{2}\right)^2 \left(\frac{1}{e}\right) \text{ (reflected torque)} \tag{4.86}$$

$$J_R = \left(\frac{W}{980.6}\right) \left(\frac{D_{PM}}{2}\right) \left(\frac{1}{e}\right) \text{ (load reflected as inertia)} \tag{4.87}$$

$$J_T = J_M + J_{PM} + J_{PI} + J_B + J_R \text{ (total inertia)} \tag{4.88}$$

$$T_{ACC} = +J_T \theta''_M + T_R \tag{4.89}$$

$$T_{DEC} = -J_T \theta''_M + T_R \tag{4.90}$$

$$T_{RUN} = T_R \tag{4.91}$$

$$T_{SS} = (\pm (F_L - F_F) + F_G) \left(\frac{D_{PM}}{2}\right) \left(\frac{1}{e}\right) \tag{4.92}$$

For horizontal systems, $F_G = 0$

For vertical systems, use: $+F_G$ when accelerating "up" or decelerating "down"
                                         $-F_G$ when accelerating "down" or decelerating "up"

Example
In a packaging machine, a 2268 g load is to be accelerated and decelerated between 0 and 122 cm s^{-1} in 0.4 s, with a total travel of 122 cm. A timing belt and pulley drive system is to be used, in which the pulleys are 5.1 cm in diameter, 15.2 cm long and made of steel. The belt weighs 2268 g. A lubricated dove tail slide structure, with a friction coefficient of 0.10 is to

be used to support the load. The motor pulley will be integral to the motor shaft, eliminating the need for a coupling.

$$J_{PM} = \pi\,(7.81)\,(15.2)\,(2.54^4)/(2 \times 980.6) = 7.92 \text{ g cm s}^2$$

$$J_{PI} = 7.92 \text{ g cm s}^2$$

$$J_B = (2268/980.6)\,(2.54^2)\,(1/0.97) = 15.5 \text{ g cm s}^2$$

$$J_R = (22\,680/980.6)\,(2.54^2)\,(1/0.97) = 155 \text{ g cm s}^2$$

$$\theta'_M = (122 \text{ cm s}^{-1})\,(\text{rev}/5.1\pi \text{ cm}) = 457 \text{ rpm} = 48 \text{ rad s}^{-1}$$

$$\theta''_M = 48/0.4 = 120 \text{ rad sec}^{-2}$$

$$F_F = (0.1)\,(22\,680) = 2268 \text{ g}$$

$$T_R = (2268)\,(5.1/2)\,(1/0.97) = 5962 \text{ g cm}$$

$$\text{Initial inertia} = 7.92 + 7.92 + 15.5 + 155 = 186 \text{ g cm s}^2$$

$$\text{Initial } T_{ACC} = (186)(120) + 5962 = 28\,282 \text{ g cm}$$

First Motor Choice: A NEMA 23 brushless servo motor with the following specs:

$$T_{CONT} = 11\,448 \text{ g cm} \quad T_{PEAK} = 34\,272 \text{ g cm}$$

$$N_{MAX} = 5000 \text{ rpm} \quad R = 7.72 \ \Omega$$

$$K_T = 1691 \text{ g cm A}^{-1} \quad J_M = 0.173 \text{ g cm s}^2$$

$$T_{ACC} = +(0.173 + 7.92 + 7.92)(120) + (15.5 + 155)(120) + 5962 = +28\,344 \text{ g cm}$$

$$T_{DEC} = -(0.173 + 7.92 + 7.92)(120) - (15.5 + 155)(120) + 5962 = -16\,420 \text{ g cm}$$

$$T_{RUN} = 5962 \text{ g cm}$$

$$T_{SS} = 0$$

In order to determine T_{RMS}, the run time must be calculated from the specifications, that is, total travel is 122 cm, run velocity is 122 cm s^{-1} and acceleration and deceleration times are 0.4 s, therefore:

$$(0.5)(0.4)(122)(2) + (122)\,(t_{RUN}) = 122$$

$$t_{RUN} = 0.6 \text{ s}$$

$$T_{RMS} = \sqrt{\frac{28344^2(0.4) + 5962^2\,(0.6) + 16420^2(0.4)}{0.4 + 0.6 + 0.4}} = 17\,938 \text{ g cm}$$

This RMS torque is 50% larger than the continuous torque rating of the motor and therefore the motor is too small for this application even though the peak torque rating is larger than the acceleration and deceleration torques.

A new selection must be made based on the RMS torque requirement.

Second Motor Choice: A NEMA 34 brushless servo motor with the following specs:

$$T_{CONT} = 406 \text{ oz in} \qquad T_{PEAK} = 1217 \text{ oz in}$$
$$N_{MAX} = 4500 \text{ rpm} \qquad R = 1.7 \, \Omega$$
$$K_T = 59.23 \text{ oz in A}^{-1} \qquad J_M = 0.0053 \text{ oz in s}^2$$

Although this selection has twice the inertia of the first choice, it is a small percentage of the total inertia and the torques will essentially remain the same as initially calculated. This motor can therefore be used for the application.

$$I_{ACC} = 28\,344/4265 = 6.7 \text{ A} \quad I_{DEC} = 16\,420/4265 = 3.8 \text{ A}$$
$$I_{RUN} = 5962/4265 = 1.4 \text{ A} \quad I_{RMS} = 17\,938/4265 = 4.2 \text{ A}$$

4.9.6 Linear Motion – Rack and Pinion Drive

Rack and pinion systems typically have the motor + pinion stationary, with the rack and load moving; however, for some designs, especially those with long travel, it is often more expeditious to have the rack remain stationary and have the motor + pinion plus load mounted together and move (see Figure 4.79).

System Data:

$$J_M = \text{motor inertia (g cm s}^2)$$
$$J_P = \text{pinion inertia (g cm s}^2)$$
$$W = \text{load weight (g)}$$
$$F_L = \text{load force (g)}$$
$$F_G = \text{gravity force} = W(g)$$

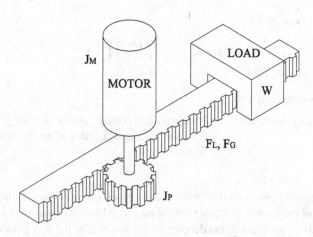

Figure 4.79 Linear motion; rack and pinion drive

$$F_F = \text{friction force} = \mu W(\text{g})$$

$$e = \text{system efficiency}$$

$$\mu = \text{friction coefficient}$$

Note: For the case in which the rack moves, the rack weight must be included in W.
For the case in which the motor moves, the motor + pinion weight must be included in W.
Motion:

$$\text{Position: } \theta_M = \frac{S}{D_P/2} \ (\text{rad})$$

$$\text{Velocity: } \theta'_M = \frac{S'}{D_P/2} \ (\text{rad s}^{-1})$$

$$\text{Acc/Dec: } \theta''_M = \frac{S''}{D_P/2} \ (\text{rad s}^{-2})$$

At Motor:

$$T_R = (\pm F_L + F_F \pm F_G)\left(\frac{D_P}{2}\right)\left(\frac{1}{e}\right) \quad \text{reflected torque} \tag{4.93}$$

$$J_R = \left(\frac{W}{980.6}\right)\left(\frac{D_P}{2}\right)^2\left(\frac{1}{e}\right) \quad \text{(load reflected as inertia)} \tag{4.94}$$

$$J_T = J_M + J_P + J_R \text{ (total inertia)} \tag{4.95}$$

$$T_{ACC} = +J_T\theta''_M + T_R \tag{4.96}$$

$$T_{DEC} = -J_T\theta''_M + T_R \tag{4.97}$$

$$T_{RUN} = T_R \tag{4.98}$$

$$T_{SS} = (\pm(F_L - F_F) + F_G)\left(\frac{D_P}{2}\right)\left(\frac{1}{e}\right) \tag{4.99}$$

For horizontal systems, $F_G = 0$

For vertical systems, use: $+F_G$ when accelerating "up" or decelerating "down"
$-F_G$ when accelerating "down" or decelerating "up"

Example
A 68 040 g table is to move a total of 396 cm at a velocity of 305 cm s^{-1}. during the constant velocity portion of the profile. The table will be mounted on roller bearing slides with a friction coefficient of 0.01. Acceleration and deceleration times will be 0.3 s. A pinion with a working diameter of 8 cm together with a rack rated at 224 532 g of dynamic thrust has been selected

for preliminary consideration.

$$J_P = 26 \text{ g cm s}^2 \text{ (catalog data)}$$

$$J_R = \left(\frac{68\,040}{980.6}\right)(8/2)^2\,(1/0.97) = 1145 \text{ g cm s}^2$$

$$\theta'_M = \left(\frac{305 \times 2\pi}{\pi \times 8}\right) = 76.3 \text{ rad s}^{-1} = 728 \text{ rpm}$$

$$\theta''_M = 76.3/0.3 = 254 \text{ rad sec}^{-2}$$

$$F_F = 0.01 \times 68\,040 = 680 \text{ g}$$

$$T_F = (680)\,(8/2)\,(1/0.97) = 2806 \text{ g}$$

$$\text{Initial inertia} = 26 + 1145 = 1171 \text{ g cm s}^2$$

$$\text{Initial } T_{ACC} = (1171)\,(254) + 2806 = 300\,240 \text{ g cm}$$

Motor Choice: A size 142 brushless servo motor with the following specs:

$$T_{CONT} = 230\,400 \text{ g cm} \quad T_{PEAK} = 691\,200 \text{ g cm}$$

$$N_{MAX} = 3000 \text{ rpm} \quad R = 0.34 \text{ }\Omega$$

$$K_T = 11\,520 \text{ g cm A}^{-1} \quad J_M = 22.8 \text{ g cm s}^2$$

$$T_{ACC} = +(22.8 + 26)(254) + (1145)(254) + 2806 = 306\,031 \text{ g cm}$$

$$T_{DEC} = -(22.8 + 26)(254) - (1145)(254) + 2806 = -300\,419 \text{ g cm}$$

$$T_{RUN} = 2806 \text{ g cm}$$

$$T_{SS} = 0$$

In order to determine T_{RMS}, the run time must be calculated from the specifications, that is, total travel is 396 cm, run velocity is 305 cm s^{-1} and acceleration and deceleration times are 0.3 s, therefore:

$$(0.5)(0.3)(305)(2) + (305)\,(t_{RUN}) = 396$$

$$t_{RUN} = 1 \text{ s}$$

$$T_{RMS} = \sqrt{\frac{306\,031^2(0.3) + 2806^2(1) + 300\,419^2(0.3)}{0.3 + 1 + 0.3}} = 185\,707 \text{ g cm}$$

$$I_{ACC} = 306\,031/11\,520 = 26.6 \text{ A} \quad I_{DEC} = 300\,419/11\,520 = 26 \text{ A}$$

$$I_{RUN} = 2806/11\,520 = 0.25 \text{ A} \quad I_{RMS} = 185\,707/11\,520 = 16 \text{ A}$$

4.9.7 Linear Motion – Roll Feed Drive

In designing a roll feed system, four items that must be considered are:

1. The unsupported weight of the material being moved must be included with the inertia of the rollers.
2. The spool of material from which the material is being removed may have to be accelerated by the roll feed.
3. The force which is used to load the two rollers together can cause an increase in the bearing torque above the normal free wheeling bearing torque.
4. Due to mounting tolerances and material compression, this force will not be perfectly normal to the center line of the roller bearing axis, and will create an additional load torque.

See Figure 4.80.
  System Data:

$$J_M = \text{motor inertia (g cm s}^2)$$

$$J_{RM} = \text{motor roller inertia (g cm s}^2)$$

$$J_{RP} = \text{pinch roller inertia (g cm s}^2)$$

$$W = \text{load weight (g)}$$

$$F_L = \text{load force (g)}$$

$$F_F = \text{friction force (g)}$$

$$T_P = \text{pressure torque (g cm)}$$

$$T_B = \text{bearing torque (g cm)}$$

$$D_{RM} = \text{motor roller diameter (cm)}$$

$$J_{SUPPLY} = \text{supply reel inertia (g cm s}^2)$$

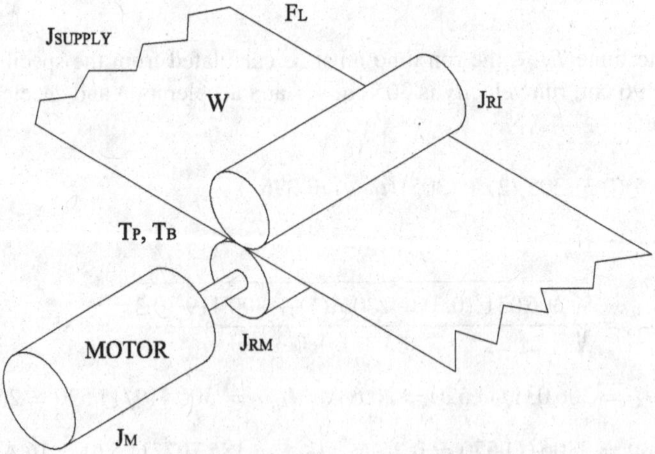

Figure 4.80   Linear motion; roll feed drive

Material weight reflected to motor as an equivalent inertia:

$$J_W = \left(\frac{W}{980.6}\right)\left(\frac{D_{RM}}{2}\right)\left(\frac{1}{e}\right)$$

Motion:

$$\text{Position: } \theta_M = \frac{S}{D_{RM}/2} \text{ (rad)}$$

$$\text{Velocity: } \theta'_M = \frac{S'}{D_{RM}/2} \text{ (rad s}^{-1}\text{)}$$

$$\text{Acc/Dec: } \theta''_M = \frac{S''}{D_{RM}/2} \text{ (rad s}^{-2}\text{)}$$

At Motor:

$$T_R = (F_L + F_F)\left(\frac{D_{RM}}{2}\right)\left(\frac{1}{e}\right) + T_P + T_B \tag{4.100}$$

$$J_T = J_M + J_{RM} + J_{RP} + J_W + J_{SUPPLY} \text{ (total inertia)} \tag{4.101}$$

$$T_{ACC} = +J_T\theta''_M + T_R \tag{4.102}$$

$$T_{DEC} = -J_T\theta''_M + T_R \tag{4.103}$$

$$T_{RUN} = T_R \tag{4.104}$$

$$T_{SS} = (F_L - F_F)\left(\frac{D_{RM}}{2}\right)\left(\frac{1}{e}\right) + T_P - T_B \tag{4.105}$$

Example

A roll feed system is to be used to uncoil 30 480 cm feet of a thin film, 17.8 cm wide from a 15.2 cm diameter full reel weighing 2268 g at 61 cm s^{-1}. A hysteresis brake will be back driven to create 907 g of tension in the film. The capstan will be a rubber-coated aluminum cylinder 5.1 cm in diameter and 20.3 cm long. The pinch roll will be an aluminum cylinder 2.54 cm in diameter and 20.3 cm long. A force of 2268 g will be used to press the pinch roller against the capstan. It is anticipated that the 2268 g force will be off center no more than 0.318 cm. Total bearing torque is expected to be 432 g cm maximum. 30.5 cm of the film will be suspended between the supply roll and the roll feed system

$$J_{RM} = \frac{\pi(2.66)(20.3)(5.1/2)^4}{2 \times 980.6} = 3.6 \text{ g cm s}^2$$

$$J_{RP} = \frac{\pi(2.66)(20.3)(2.54/2)^4}{2 \times 980.6} = 0.22 \text{ g cm s}^2$$

$$J_{SUPPLY\ REEL} = (2268/2 \times 980.6)\left(7.62^2\right) = 67.1 \text{ g cm s}^2$$

$$T_P = (2268)(0.318) = 721 \text{ g cm}$$

$$T_B = 432 \text{ g cm}$$

$$F_L = 907 \text{ g}$$

$$F_F = 0(\text{the film is suspended between the supply reel and the roll feed})$$

$$T_R = (907)(5.1/2) + 721 + 432 = 3466 \text{ g cm}$$

$$W = (30.5/30\,480)(2268) = 2.3 \text{ g (weight of the 30.5 cm of film in suspension)}$$

$$J_W = (2.3/980.6)(5.1/2)^2 = 0.015 \text{ g cm s}^2 \text{ (can be neglected)}$$

Assume accelerate time $= 1$ s

$$S' = 61 \text{ cm s}^{-1}$$

$$S'' = 61 \text{ cm s}^{-2}$$

$$\theta_M' = \left(61 \text{ cm s}^{-1}\right)(1 \text{ rev}/\pi \times 5.1 \text{ cm}) = 3.81 \text{ rev s}^{-1} = 24 \text{ rad s}^{-1} = 230 \text{ rpm}$$

$$\theta_M'' = 24 \text{ rad s}^{-2}$$

Initial inertia $= 3.6 + 0.22 + 67.1 = 71 \text{ g cm s}^2$

Initial $T_{ACC} = (71)(24) + 3466 = 5179 \text{ g cm}$

First Motor Choice: A size 16 brushless servo motor with the following specs:

$$T_{CONT} = 4176 \text{ g cm} \quad T_{PEAK} = 12\,456 \text{ g cm}$$

$$N_{MAX} = 5000 \text{ rpm} \quad R = 4.65 \ \Omega$$

$$K_T = 1574 \text{ g cm A}^{-1} \quad J_M = 0.036 \text{ g cm s}^2$$

$$T_{ACC} = +(0.036 + 3.6 + 0.22 + 67.1)(24) + 3466 = 5169 \text{ g cm}$$

$$T_{DEC} = -(0.036 + 3.6 + 0.22 + 67.1)(24) + 3466 = 1763 \text{ g cm}$$

$$T_{RUN} = 347 \text{ g cm}$$

$$T_{SS} = 0$$

T_{RMS} need not be calculated since it will take 8.3 min to unload the complete reel and therefore $T_{RMS} \approx T_{RUN}$

Note: The motor always supplies a "+" torque; during deceleration, the hysteresis brake, working against the motor, will cause the system to decelerate.

The motor will be operating far below its rated speed and, therefore, inefficiently.

A better approach would be to use a smaller motor combined with a 10:1 gearhead as follows.

Second motor choice: A size 16 brushless servo motor with the following specs:

$$T_{CONT} = 1512 \text{ g cm} \quad T_{PEAK} = 4608 \text{ g cm}$$

$$N_{MAX} = 5000 \text{ rpm} \quad R = 4.31 \ \Omega$$

$$K_T = 621 \text{ g cm A}^{-1} \quad J_M = 0.013 \text{ g cm s}^2$$

Plus a matching size 16 gearhead with a 10:1 ratio

$$\theta'_M = 240 \text{ rad s}^{-1} = 2300 \text{ rpm}$$

$$\theta''_M = 240 \text{ rad s}^{-2}$$

$$T_{ACC} = +(0.013 + (3.6 + 0.22 + 67.1)(1/100))(240) + 3466/10 = +520 \text{ g cm}$$

$$T_{DEC} = -(0.013 + (3.6 + 0.22 + 67.1)(1/100))(240) + 3466/10 = +173 \text{ g cm}$$

$$T_{RUN} = 347 \text{ g cm}$$

$$T_{SS} = 0$$

$$I_{ACC} = 520/621 = 0.84 \text{ A} \quad I_{DEC} = 173/621 = 0.3 \text{ A}$$

$$I_{RUN} = 347/621 = 0.6 \text{ A} \quad I_{RMS} = 0.6 \text{ A}$$

4.9.8 Linear Motion – Linear Motor Drive

The latest linear motion mechanism, made practical and economically feasible by the advent of brushless motor technology, is the linear motor. It is essentially the linear dual of the rotary brushless motor and creates linear motion directly without the need for motion conversion mechanisms. As such, the basic equations are simply those determining the necessary forces required for the four parts of the profile. See Figure 4.81.

System Data:

$$W = \text{load weight (g)}$$

$$W_M = \text{motor weight (g)}$$

$$F_L = \text{load force (g)}$$

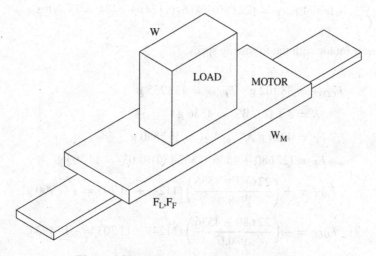

Figure 4.81 Linear motion; linear motor drive

$$F_F = \text{friction force (g)}$$
$$F_{MA} = \text{magnetic attraction force (iron core motors) (g)}$$
$$\mu = \text{friction coefficient}$$

At Motor:

$$F_F = \mu(W + W_M + F_{MA}) \qquad (4.106)$$

$$F_{ACC} = + \left(\frac{W + W_M}{980.6} \right) S'' \pm F_L + F_F \qquad (4.107)$$

$$F_{DEC} = - \left(\frac{W + W_M}{980.6} \right) S'' \pm F_L + F_F \qquad (4.108)$$

$$F_{RUN} = \pm F_L + F_F \qquad (4.109)$$

$$F_{SS} = \pm(F_L - F_F) \text{ if } F_L > F_F \qquad (4.110)$$

$$= 0 \text{ if } F_L < F_F$$

Example
A 22 680 g load is to be oscillated back and forth over a 91 cm stroke with no dwell time.
Acceleration and deceleration are to be in 0.17 to and from a run velocity of 191 cm s^{-1} in a trapezoidal profile.
For initial calculations, assume that F_{MA} equals 226 800.

$$\text{Initial } F_F = (22\,680 + 226\,800)(0.03) = 7484 \text{ g}$$
$$S'' = 191/0.17 = 1124 \text{ cm s}^{-2}$$
$$\text{Initial } F_{ACC} = (22\,680/980.6)(1124) + 7484 = 33\,481 \text{ g}$$

Select a linear motor with the following specs:

$$F_{CONT} = 38\,102 \text{ g} \quad F_{PEAK} = 139\,255 \text{ g}$$
$$R = 4.1\ \Omega \quad W_M = 4536 \text{ g}$$
$$F_C = 4854 \text{ g A}^{-1} \quad F_{MA} = 362\,880 \text{ g}$$
$$F_F = (22\,680 + 4536 + 362\,880)(0.03) = 11\,703 \text{ g}$$
$$F_{ACC} = + \left(\frac{22\,680 + 4536}{980.6} \right)(1124) + 11\,703 = +42\,900 \text{ g}$$
$$F_{DEC} = - \left(\frac{22\,680 + 4536}{980.6} \right)(1124) + 11\,703 = -19\,493 \text{ g}$$
$$F_{RUN} = 11\,703 \text{ g}$$

Calculate the run time in order to calculate F_{RMS}

$$(0.5)(0.17)(191)(2) + (191)(t_{RUN}) = 91$$

$$t_{RUN} = 0.3 \text{ s}$$

$$F_{RMS} = \sqrt{\frac{42\,900^2(0.17) + 11\,703^2(0.3) + 19\,493^2(0.17)}{0.17 + 0.3 + 0.17}} = 25\,573 \text{ g}$$

$$I_{ACC} = 42\,900/4854 = 8.8 \text{ A} \quad I_{DEC} = 19\,493/4854 = 4 \text{ A}$$

$$I_{RUN} = 11\,703/4854 = 2.4 \text{ A} \quad I_{RMS} = 25\,573/4854 = 5.3 \text{ A}$$

References

[1] Kou, B. and Tal, J. (1978) *DC Motors and Control Systems*, SRL Pub Co., Champaign, IL, pp. 236–242.

[2] Hausman, E. and Slack, E.P. (1951) *Physics*, D. Van Nostrand, N.Y., p. 70.

[3] Hansen, D. (Dec. 2010) Direct Drives for Inertia Matching, Design News, pp. 46–49.

[4] Parker Corp., Max Plus – J (MPI) Motor Line.

[5] Stephans, L. (2004) What goes wrong when inertias aren't right, Machine Design.

[6] Waagen, H. (April, 1969) Reduce Torsional Resonance in Incremental Servo, Control Engineering.

[7] Kou, B. and Tal, J. (1978) *DC Motors and Control Systems*, SRL Pub Co., Champaign IL, pp. 113–123.

[8] Chestnut, H. and Mayer, R. (1951) *Servomechanisms and Regulating System Design*, John Wiley & Sons, Inc., New York, pp. 129–130.

[9] Chestnut, H. and Mayer, R. (1951) *Servomechanisms and Regulating System Design*, John Wiley & Sons, Inc., New York, p. 134.

[10] Astrom, K. and Hagglund, T. (1995) *PID Controllers: Theory, Design and Tuning*, Instrument Society of America, Research Triangle Park, NC, pp. 64–70.

[11] Kou, B. and Tal, J. (1978) *DC Motors and Control Systems*, SRL Pub Co., Champaign, IL, pp 72–75.

[12] Raskin, C. (Sept./Oct., 1997) The Science of Tuning Servo Motors, Motion Control, pp. 24–29.

[13] Ellis, G. (1991) *Control System Design Guide*, Academic Press, San Diego, CA.

[14] Kou, B. and Tal, J., (1978) *DC Motors and Control Systems*, SRL Pub Co., Champaign, IL, pp. 17–19.

5

System Examples – Design and Simulation

5.1 Linear Motor Drive

In this application, the primary requirement is to accelerate an 18 kg (176.5 N) load, including the weight of the carriage, at a minimum of 1 g (9.806 m s^{-2}) to reach a velocity of 0.1 m s^{-1} within 10 ms after the start command.

Slew times will be variable, with a minimum of 0.19 s to result in a move distance of 0.02 m (2 cm).

To accommodate rapid, short moves, a three phase linear DC motor drive has been selected to provide smooth and vibration free operation.

Initial force calculation:

$$F = ma = \frac{W}{g}g = 176.5\,\text{N} \tag{5.1}$$

Evaluate a model ICD10-050A1 with the following specifications:

$K_F = 89.2\,\text{N A}^{-1}$	$K_E = 72.8\,\text{V m}^{-1}\,\text{s}^{-1}$
$R = 9\,\Omega$	$L = 29\,\text{mH}$
Peak Current $= 7.9\,\text{A}$	Continuous Current $= 1.9\,\text{A}$ @ 130 °C coil temperature
M_m (coil assembly mass)	1.9 Kg
F_a (magnetic attraction)	1780 N
R_{th} (thermal resistance)	1.52 °C W^{-1}

Electromechanical Motion Systems: Design and Simulation, First Edition. Frederick G. Moritz.
© 2014 John Wiley & Sons, Ltd. Published 2014 by John Wiley & Sons, Ltd.
Companion Website: www.wiley.com/go/moritz

In addition to the force needed to achieve the required acceleration of the load inertia, there will be a frictional force determined by the coefficient of friction and three components:

- The load weight
- The coil assembly mass M_m
- The magnetic attraction F_a.

Assuming a coefficient of friction of 0.03, the frictional force:

$$F_F = 0.03\,(176.5 + 1.9 \times 9.807 + 1780) = 59.3\,\text{N} \tag{5.2}$$

Forces and currents will then be:

$$F_{ACC} = 176.5 + 59.3 = 235.8\,\text{N} \qquad I_{ACC} = \frac{235.8}{89.2} = 2.6\,\text{A} \tag{5.3}$$

$$F_{DEC} = -176.5 + 59.3 = -117.2\,\text{N} \quad I_{DEC} = \frac{-117.2}{89.2} = -1.3\,\text{A} \tag{5.4}$$

$$F_{RUN} = 59.3\,\text{N} \quad I_{RUN} = \frac{59.3}{89.2} = 0.7\,\text{A} \tag{5.5}$$

Maximum RMS current, assuming the short move is repeated continually without a pause between moves:

$$I_{RMS} = \sqrt{\frac{(2.6^2)\,(0.01) + (0.7^2)\,(0.19) + (1.3^2)\,(0.01)}{0.21}} = 0.92\,\text{A} \tag{5.6}$$

Dissipation:

$$P = I_{RMS}^2 R = \left(\frac{3}{4}\right)(0.92^2)\,(9) = 5.7\,\text{W} \tag{5.7}$$

Coil temperature rise:

$$Temp_{RISE} = PR_{th} = 5.7 \times 1.52 = 9\,^\circ\text{C} \tag{5.8}$$

Based on Equations 5.7 and 5.8, the motor will be operated well within its thermal limits and no further thermal calculations are required.

Maximum voltage:

$$E_{MAX} = I_{ACC}R + S'_{MAX}K_E = 2.6 \times 9 + 0.1 \times 72.8 = 31\,\text{V} \tag{5.9}$$

A simulation of the system is shown in Figure 5.1.

A linear encoder with a resolution of 75 lines per cm creating a resolution of 300 pulses per cm after an X4 and an overall resolution of 30 000 pulses per m will be used.

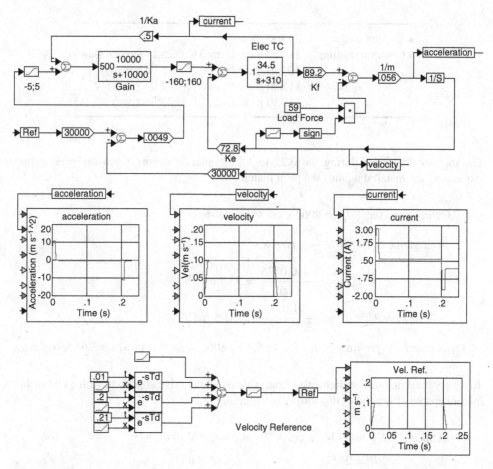

Figure 5.1 Linear motor drive simulation (available in full color at www.wiley.com/go/moritz)

The digital signals will pass through a D to A 12 bit circuit, operating from a ±10 V supply, resulting in a gain of 0.0049.

The velocity reference command has been created with four delayed ramp generators with positive and negative slopes summed to produce the reference as shown.

The current amplifier contains a 1600 Hz low pass filter to eliminate oscillation bursts that occurred on the leading and trailing edges of the current pulses.

The current amplifier has a gain of 2 A V^{-1}, with a current limit of ±5 A and an output voltage limit of ±160 V.

5.2 Print Cylinder Control

A servo system is required to control the velocity of a print cylinder such that it reaches a velocity of 10 rpm (1.05 rad s^{-1}) within 1° (0.0175 rad) after the start of motion while being subjected to the following load torque profile.

- Stiction torque: 28 800 g cm
- Coulomb torque: 11 520 g cm from 0 to 0.25 s (acc and slew)
                    40 120 g cm from 0.25 to 1.75 s (slew)
                    11 520 g cm from 1.75 to 2.0 s (slew)
                    11 250 g cm from 2.0 to 2.25 s (dec)

The increase in torque during the 0.25 to 1.75 s interval occurs when the print cylinder "squeezes" the material against the print platen.

Cylinder inertia: $J_C = 8064$ g cm s^2 (calculated)

Determine t_{ACC}: $\theta = \dfrac{1}{2} \times t_{ACC} \times \theta'$

$$t_{ACC} = \frac{2 \times 0.0175}{1.05} = 33.3 \text{ m s}$$

Determine θ'': $\theta'' = \dfrac{1.05}{33.3 \times 10^{-3}} = 31.5 \text{ rad s}^{-2}$

Determine $T_{ACC}(\text{prelim.}) = J_C\theta'' + 11\,520 = 8064 \times 31.5 + 11\,520 = 265\,536$ g cm

To satisfy the dynamic requirements of the system a direct drive approach, with a kit brushless motor capable of supplying this torque is required.

Motor Specs.: $K_T = 79\,142$ g cm A^{-1} $K_E = 7.8$ V rad^{-1} s^{-1}

$R = 25\,\Omega$ $L = 204$ mH

$J_M = 14.4$ g cm s^2

$T_{ACC} = (J_C + J_M)\theta'' + 11\,520 = 8078 \times 31.5 + 11\,520 = 265\,980$ g cm

$$I_{ACC} = \frac{T_{ACC}}{K_T} = \frac{265\,980}{79\,142} = 3.4 \text{ A}$$

$E_{PEAK} = I_{ACC}R + K_E\theta' = 3.4 \times 25 + 7.8 \times 1.05 = 93$ V

A 160 V, 5 A brushless servo amplifier will meet the system requirements
Determine I_{RMS}

Current Profile: $I = 3.4$ A from 0 to 0.033 s (acc and slew)

$= 11\,520/79\,142 = 0.15$ A from 0.033 to 0.25 s (slew)

$= 40\,320/79\,142 = 0.51$ A from 0.25 to 1.75 s (slew)

$= 11\,520/79\,142 = 0.15$ A from 1.75 to 2 s (slew)

$$I_{DEC}: \theta'' = \frac{1.05 \text{ rad s}^{-1}}{0.25 \text{ s}} = 4.2 \text{ rad s}^{-2}$$

$$T_{DEC} = 8078 \times 4.2 - 11520 = 22\,408 \text{ g cm}$$

$$I_{DEC} = \frac{22\,408}{79\,142} = 0.3 \text{ A from 2 to 2.25 s}$$

$$I_{RMS} = \sqrt{\frac{3.4^2 \times 0.033 + 0.15^2 \times 0.217 + 0.51^2 \times 1.5 + 0.15^2 \times 0.25 + 0.3^2 \times 0.25}{2.25}}$$

$$= 0.6 \text{ A}$$

Motor dissipation: $= \frac{3}{4} I^2 R = 6.8 \text{ W (for 3 phase brushless motor)}$

Encoder: The precise motion and print control requires the use of a special high resolution 590 000 count encoder. With a X4 multiplier this will provide 2 360 000 counts per revolution equal to 375 605 counts per radian.

Max encoder freq.: 590 000 counts rev^{-1} × 10 rpm ×1 min/60 s = 98.3 kHz

This is well below the maximum rating of 150 kHz.

A 16 bit DAC will be used with an output range of ±10 V, resulting in a gain of:

$$G_{DAC} = \frac{20}{2^{16}} = 0.00031 \text{ V/bit}$$

Figure 5.2 shows the system simulation.

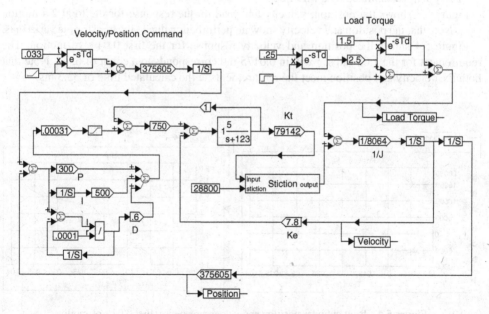

Figure 5.2 Print cylinder control simulation

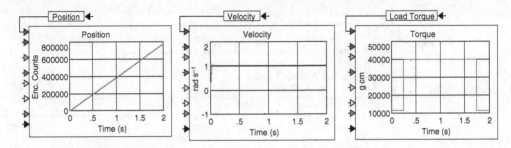

Figure 5.3   Print cylinder position, velocity and load torque response

The command block consists of a velocity ramp generator with a slope of 31.5 and leveled at 0.033 second to command the slew velocity of 1.05 rad s^{-1}. The velocity signal is calibrated into encoder counts and converted into the position command by integration.

The developed torque passes through a stiction block (see Section 4.8.2) and is summed with the load torque before being applied to the inertia.

The load torque is created per the coulomb torque listing described previously. An initial value of 11 520 is increased to 40 320 by multiplying the 11 520 value by 2.5 at 0.25 s and adding it to the 11 520 value for a total of 40 320. At 1.75 s the multiplied value is then subtracted, reducing the torque to 11 520.

A transconductance amplifier with a nominal DC gain of 1 A V^{-1} drives the motor. For this critical application the motor should be of the sinusoidal BEMF type driven by a sinusoidal amplifier.

The system is stabilized by a PID filter.

Figure 5.3 shows the position, velocity and load torque response for the total 2 s motion time. Note that the position and velocity show no perturbation due to the load torque variations.

Figure 5.4 shows the position and velocity response for the first 0.04 s of motion. The requirement for the 1.05 rad s^{-1} within 0.0175 rad corresponds to a count of 6564. Note that both the velocity and position meet the requirements at the calculated time of 33.3 ms.

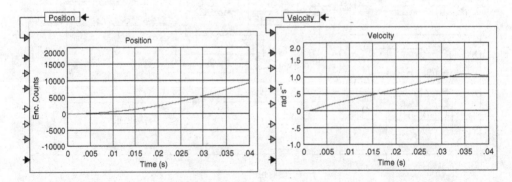

Figure 5.4   Print cylinder position and velocity response, first 0.04 s of motion

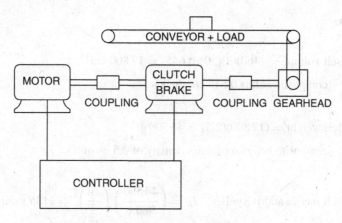

Figure 5.5 Conveyor system, clutch/brake control diagram

5.3 Conveyor System – Clutch/Brake Control

A conveyor system is being designed with the following specifications:

Load:	Qty 1 to 10 containers @ 34 000 g $\pm$ 10% each
Slew Velocity:	90 cm s^{-1}
Travel:	460 cm
Pulleys:	10 cm diameter, 30 cm long, steel
Motor Velocity:	1750 rpm nominal
Acc Time:	0.5–1 s
Control:	Clutch/Brake

The system diagram is shown in Figure 5.5
   Object:

1. Determine the motor, clutch and brake ratings
2. Simulate the system to demonstrate dynamic performance.

5.3.1 Determine Basic Parameters

- Pulley inertia:

$$J_P = \frac{\pi (7.75)(30)(5^4)}{2(980.6)} = 233 \ \mathrm{g\,cm\,s^2} \tag{5.10}$$

- Belt Inertia:

$$\text{Belt volume} = (460)(2)(30)(0.645) = 17\,800 \text{ cm}^3$$

  (composite belt is 0.645 cm thick)

$$\text{Belt weight} = (17\,800)(2.5) = 44\,500 \text{ g}$$

  (composite belt material has density of 2.5 g cm^{-3})   (5.11)

$$\text{Belt inertia at drive pulley, } J_B = \left(\frac{44\,500}{980.6}\right)\left(\frac{10}{2}\right)^2 = 1135 \text{ g cm s}^2 \quad (5.12)$$

- Load inertia at drive pulley,

$$J_{L(\min)} = \left(\frac{34\,000 \times 0.9}{980.6}\right)\left(\frac{10}{2}\right)^2 = 780 \text{ g cm s}^2 \tag{5.13}$$

$$J_{L(\max)} = \left(\frac{340\,000 \times 1.1}{980.6}\right)\left(\frac{10}{2}\right)^2 \times 9534 \text{ g cm s}^2 \tag{5.14}$$

- Gearhead Ratio:

$$\text{Pulley velocity, } N_P = \frac{90 \text{ cm}}{s} \times \frac{60 \text{ s}}{\min} \times \frac{\text{rev}}{\pi \times 10 \text{ cm}} = 172 \text{ rpm} = 18 \text{ rad s}^{-1}$$
$$\tag{5.15}$$

  Motor velocity, $N_M = 1750$ rpm

$$\text{Gearhead ratio, } N = \frac{1750}{172} = 10 : 1 \tag{5.16}$$

- Friction:

$$\text{Assume support structure friction coefficient, } \mu = 0.15$$

$$\text{Friction} = 34\,000 \times 1.1 \times 10 \times 0.15 = 56\,100 \text{ g} \tag{5.17}$$

$$T_F = 56\,100 \times 5 = 280\,500 \text{ g cm} \tag{5.18}$$

- Gearhead efficiency, e: assume 0.8.

- Total inertia at gearhead output,

$$J_T = J_{L(\text{max})} + J_B + J_P + J_P$$

$$= 9534 + 1135 + 233 + 233 = 11\,135 \text{ g cm s}^2 \tag{5.19}$$

- Total inertia at gearhead input,

$$J_{IN} = \frac{J_T}{N^2}$$

$$= \frac{11\,135}{{}^{\prime}10^2} = 111.4 \text{ g cm s}^2 + \text{gearhead} \tag{5.20}$$

- Load Acceleration,

$$\theta_L'' = \frac{18}{1} \text{ to } \frac{18}{0.5} = 18 \text{ to } 36 \text{ rad s}^{-2} \tag{5.21}$$

- Gearhead input acceleration,

$$\theta_{IN}'' = N\theta_L'' = 180 \text{ to } 360 \text{ rad s}^{-2} \tag{5.22}$$

- Maximum torque at gearhead output,

$$T_{out(\text{max})} = (J_T)(\theta_L'') + T_F$$

$$= (11\,135)(36) + 280\,500 \tag{5.23}$$

$$= 681\,400 \text{ g cm} = 67 \text{ N m}$$

- Maximum torque at gearhead input,

$$T_{IN(\text{max})} = \frac{T_{out(\text{max})}}{N \times e}$$

$$= \frac{681\,400}{10 \times 0.8} = 85\,200 \text{ g cm} \tag{5.24}$$

$$\text{Horsepower (preliminary)} = \frac{85\,200 \times 1750}{7.26 \times 10^7} = 2 \text{ HP} \tag{5.25}$$

5.3.2   Initial Component Selection

Motor:		2 HP @ 1750 rpm
Gearhead:	Size 115	$T = 147\,\text{N m} = 1.5 \times 10^6\,\text{g cm}$
		Velocity = 2900 rpm (nom), 4500 rpm (max)
		$J_{GH} = 1.12\,\text{g cm s}^2$
Clutch/Brake:	Size 180	$T_{(dynamic)} = 207\,400\,\text{g cm @ 1800 rpm}$
		$J_{GB} = 10\,\text{g cm s}^2$

$$
\begin{aligned}
\text{Torque load on motor, } T_L &= \frac{(J_{IN} + J_{GH})(\theta''_{IN})}{e} + (J_{CB})(\theta''_{IN}) + \frac{T_F}{N \times e} \\
&= \frac{(111.4 + 1.12)(360)}{0.8} + (10)(360) + \frac{280\,500}{10 \times 0.8} \quad (5.26) \\
&= 54\,234 + 35\,063 \\
&= 89\,300\,\text{g cm}
\end{aligned}
$$

$$
\text{Horsepower} = \frac{89\,300 \times 1750}{7.26 \times 10^7} = 2.2\,\text{HP} \quad \text{Use a 2.5 or 3 HP motor.} \quad (5.27)
$$

5.3.3   Simulate System

The motor is simulated using the "slip" model described in Section 3.1.

For a 4 pole squirrel cage motor, the synchronous speed is 1800 rpm and the slip/speed equation is:

$$
\begin{aligned}
S &= \frac{1800 - N}{1800} \text{ for } N \text{ in rpm} \\
&= \frac{188.5 - N}{188.5} = 1 - 0.0053\,N \text{ for } N \text{ in rad s}^{-1}
\end{aligned} \quad (5.28)
$$

Using a generic class A speed/torque curve for a 2.5 HP motor (Figure 5.5), based on a continuous maximum rating of 1750 rpm @ 103 680 g cm, slip versus torque data can be listed and plotted from 100 to 0% of synchronous speed (Figure 5.6).

Using Excel curve fitting, the torque versus slip equation can be derived as shown in Figure 5.7.

$$
T = -228.1S^4 + 586.06S^3 - 516.98S^2 + 170.55S + 2.7497\,\text{N m} \quad (5.29)
$$

The complete system simulation is shown in Figure 5.8.

The motor is modeled with Equation 5.29, with T multiplied by 10 200 to convert N m to g cm and inertia of 15 g cm s^2. The clutch, with a torque of 54 234 g cm and a control

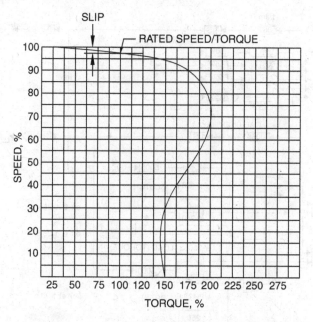

Figure 5.6 Conveyor system; motor torque/speed curve

coil delay of 20 ms is "turned on" 0.25 s after the motor reaches constant no load speed and is "turned off" at 1.25 s, with the brake "turning on" at 1.30 s to avoid clutch/brake interference.

The load torque (system friction) has two paths. During load acceleration and run it places a load on the motor. During brake, after clutch release, it assists in decelerating the load.

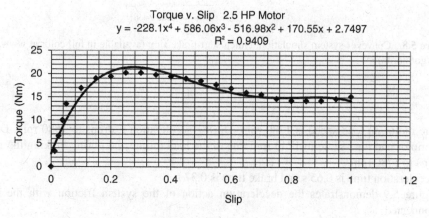

Figure 5.7 Conveyor system; torque versus slip, 2.5 HP motor

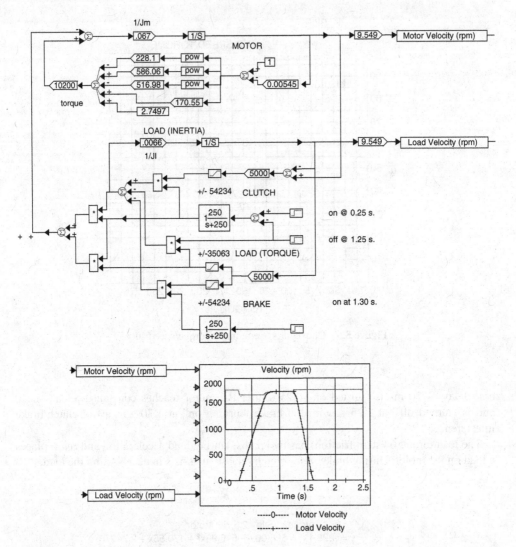

Figure 5.8 Conveyor system simulation; clutch start-brake stop (available in full color at www.wiley .com/go/moritz)

The motor no load speed is 1780 rpm. During acceleration it drops to 1680 rpm. During run, motor and load settle to 1740 rpm due to the load torque and during braking the motor returns to 1780 rpm.

Acceleration time is 0.65 s and brake time is 0.37 s.

Figure 5.9 demonstrates the deceleration action of the system friction with the brake disconnected.

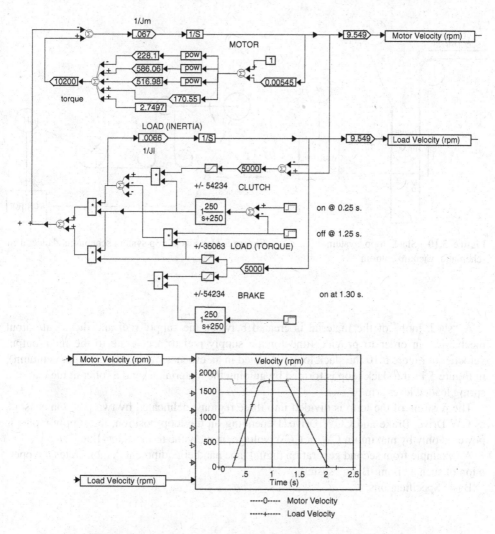

Figure 5.9 Conveyor system simulation; clutch start-friction stop

5.4 Bang-Bang Servo (Slack Loop System)

Many processes require moving a flexible material (cloth, sheet plastic, cable, etc.) between a large storage roll and a location for some form of processing.

The material is usually removed at relatively high velocity, initiated by high acceleration or deceleration while maintaining the material under constant tension.

Servoing the high inertia "supply" roll to meet these two requirements simultaneously can be difficult. A direct approach is to separate the requirements as shown schematically by the examples in Figures 5.10 and 5.11.

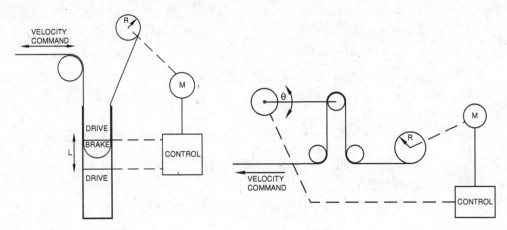

Figure 5.10   Slack loop system schematic, vacuum column

Figure 5.11   Slack loop system schematic, dancer arm

A "slack loop" of the material is created between the supply reel and the input/output mechanism in order to provide time for the supply reel to accelerate to the input/output velocity. In Figure 5.10 the slack loop is formed in an evacuated column (a vacuum column); in Figure 5.11 the slack loop is formed by guiding the material around a roller at the end of a spring loaded lever arm.

The position of the loop is divided into three regions, delineated by two position sensors, as CW Drive, Brake and CCW Drive. Depending on the loop position, the supply motor is powered only by maximum CW or CCW voltage, namely the term "Bang-Bang".

An example from second generation digital tape handler equipment demonstrates the operation of such a "Bang-Bang" system.

Basic Specifications:

Supply reel inertia full: 122.4 g cm s^2

Supply reel inertia empty: 43.2 g cm s^2

Full reel radius; 12.7 cm

Empty reel radius: 6.35 cm

Input/output velocity: 254 cm s^{-1}

Required minimum reel velocity, full: $\left(254\dfrac{\text{cm}}{\text{s}}\right)\left(\dfrac{1\ \text{rev}}{25.4\pi\ \text{cm}}\right)\left(\dfrac{2\pi\ \text{rad}}{\text{rev}}\right) = 20\dfrac{\text{rad}}{\text{s}}$

empty: $\left(254\dfrac{\text{cm}}{\text{s}}\right)\left(\dfrac{1\ \text{rev}}{12.7\pi\ \text{cm}}\right)\left(\dfrac{2\pi\ \text{rad}}{\text{rev}}\right) = 40\dfrac{\text{rad}}{\text{s}}$

Evaluate the use of the following brush motor for this application:

$$K_t = 3600\,\text{g cm A}^{-1} \quad K_e = 0.35\,\text{V rad}^{-1}\,\text{s}^{-1} \quad B = 14.4\,\text{g cm rad}^{-1}\,\text{s}^{-1}$$

$$R = 3\,\Omega \quad L = 13.5\,\text{mH}$$

$$\text{BEMF @ 20 rad s}^{-1} = 20\frac{\text{rad}}{\text{s}} \times 0.35\frac{\text{V}}{\frac{\text{rad}}{\text{s}}} = 7\,\text{V}$$

$$\text{BEMF @ 40 rad s}^{-1} = 40\frac{\text{rad}}{\text{s}} \times 0.35\frac{\text{V}}{\frac{\text{rad}}{\text{s}}} = 14\,\text{V}$$

Supply voltage selection:

In addition to operating at 254 cm s^{-1}, the equipment must operate at a rewind velocity of 508 cm s^{-1} which will result in a maximum BEMF of 28 V. During normal operation (254 cm s^{-1}) there will be an IR drop during acceleration, ranging from $\frac{E_{supply}}{R}$ at the start of acceleration to $\frac{E_{supply-BEMF}}{R}$ at the end of acceleration. Since current will only be limited by the resistance and a reasonable value of current (torque) is needed during acceleration, an initial choice of 36 V for the voltage will be evaluated. This will provide an initial acceleration current of 12 A; 43 200 g cm of torque.

A simulation of the system is shown in Figures 5.12 and 5.13.

Figure 5.12 shows the response of the full reel for an input step command of 254 cm s^{-1}.

The system operates as both a velocity and positional system. The command is summed with the reel velocity (after being multiplied by the reel radius) and divided by 2 to create the loop velocity error. The loop velocity is also integrated to create the loop position which is summed with the brake limit value (equal to $^1/_2$ the length, L, of the brake zone) of 3.75 cm to create the position error. Both errors are then summed and amplified by an arbitrarily high gain to create the bang-bang switching of the drive voltage from +36 to −36 V.

Note the following:

- The current initially rises to 12 A, then drops to 9.7 A at the end of acceleration
- The reel velocity rises to a peak of 22.5 rad s^{-1} and then settles to a *limit cycle* value of 20 ± 0.5 cm s^{-1}
- The loop Vel/Pos plot is a *phase plane* plot, with loop velocity on the vertical axis and loop position on the horizontal axis. It shows how the loop velocity initially starts at 127 cm s^{-1} at the center of the brake zone (0 cm), then decreases as the reel accelerates, exits the brake zone and reaches a peak position of 4.35 cm after which the loop drops back to the brake zone limit and settles into the limit cycle.

Figure 5.13 shows the response of the full reel to a bidirectional command at a rate of 0.7 cps.

Note how the loop velocity initially again starts at 127 cm s^{-1} and just as the loop position reaches the brake zone limit (3.75 cm), the first command reversal occurs causing the loop velocity to rapidly change to −260 cm s^{-1}.

Repeated command reversals cause the loop to settle in to peak velocity excursions of ±260 cm s^{-1} and peak loop position excursions of ±12 cm.

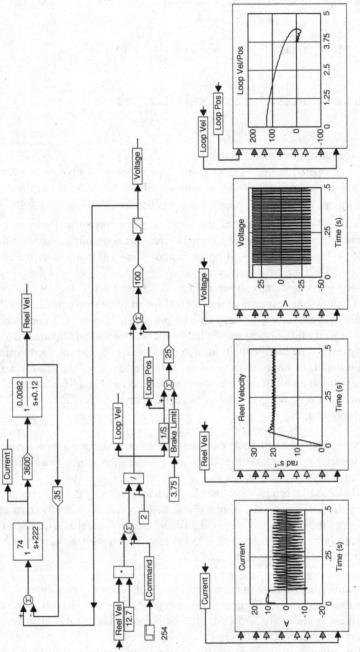

Figure 5.12 Slack loop system simulation, step command

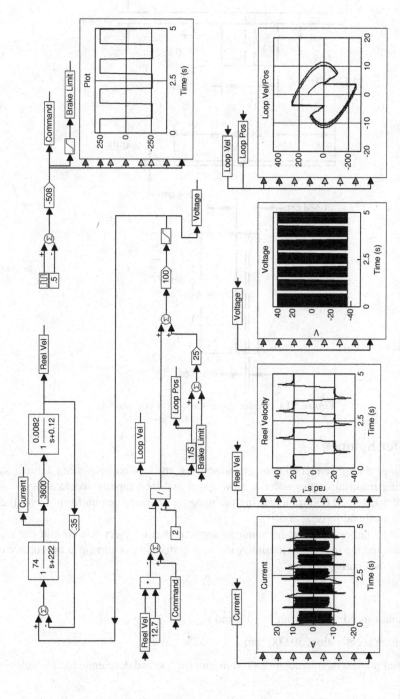

Figure 5.13 Slack loop system simulation, bidirectional command (available in full color at www.wiley.com/go/moritz)

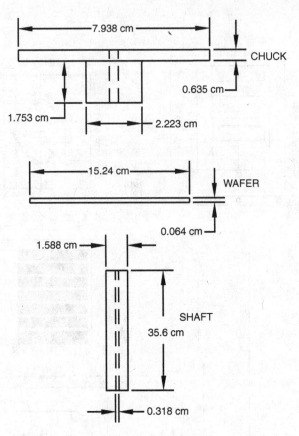

Figure 5.14   Wafer spinner assembly components

5.5   Wafer Spinner

Wafer coatings during semiconductor manufacturing are applied by placing a small stream of the coating material on the wafer at slow speed and then rapidly accelerating the wafer to speeds of 5000 to 10 000 rpm, causing the material to spread out uniformly over the disk surface by centrifugal force action.

In selecting a motor for the wafer spinning application, the object is to choose one that has the lowest dissipation in order to minimize the temperature rise occurring at the surface of the wafer during the coating process.

Specifications:

Maximum speed: 10 000 rpm $= 1047\,\text{rad s}^{-1}$

Maximum acceleration: 50 000 rpm $\text{s}^{-1} = 5236\,\text{rad s}^{-2}$

Nominal profile: Accelerate to 5000 rpm, run for 5 s, and decelerate to zero velocity.

Components: A hollow shaft connecting the motor rotor to a Delrin chuck on which the silicon wafer is held in place by vacuum introduced through the shaft (see Figure 5.14).

Inertias:

$$\text{Chuck: } J = 0.359 \, \text{g cm s}^2$$
$$\text{Wafer: } J = 0.792 \, \text{g cm s}^2$$
$$\text{Shaft: } J = 0.180 \, \text{g cm s}^2$$
$$\text{Total load inertia: } J_l = 1.33 \, \text{g cm s}^2$$
$$\text{Acc/Dec times: } 0.1 \, \text{s to } 5000 \, \text{rpm}; \, 0.2 \, \text{s to } 10\,000 \, \text{rpm}$$

At the high speeds and acceleration involved, a series of motors must be evaluated with respect to dissipation caused by I^2R during acceleration and deceleration and core loss dissipation occurring while running at 5000 and 10 000 rpm.

Based on bearing and mounting requirements, a size 90 mm motor was chosen for evaluation. Table 5.1 shows the results of a series of calculations performed to determine which motor will have the lowest dissipation for the maximum acceleration and velocities.

Table 5.1 Wafer spinner parameters versus motor size

Motor	Km	J_{motor}	P_c	J_{total}	T_{acc}	T_{rms}	I^2R	P_{core}	P_{total}	T_{rms}	I^2R	P_{core}	P_{total}
						(5 k)	(5 k)	(5 k)	(5 k)	(10 k)	(10 k)	(10 k)	(10 k)
A	490	0.046	0.37	1.38	7226	1583	10.4	4.1	14.5	2765	31.8	11.7	43.5
B	905	0.092	0.78	1.42	7435	1620	3.2	8.7	11.9	2841	9.9	24.7	34.6
C	1270	0.138	1.19	1.47	7697	1666	1.72	13.3	15	2936	5.3	37.6	42.9
D	1560	0.184	1.6	1.51	7906	1704	1.2	17.9	19.1	3011	3.7	50.6	54.3

Motor stator sizes:

$$\text{A} - 0.635 \, \text{mm}$$
$$\text{B} - 1.27 \, \text{mm}$$
$$\text{C} - 1.91 \, \text{mm}$$
$$\text{D} - 2.54 \, \text{mm}$$

Terms and Units:

K_m = motor constant; $\text{g cm}/\sqrt{w}$
J_{motor}, J_{total} = inertias; g cm s^2
P_c = core loss factor; W @ 1000 rpm
P_{core} = core loss at 5 krpm and 10 krpm; W
T_{acc} = accelerate torque for maximum acceleration; g cm
T_{rms} = torques for $t_{acc/dec} = 0.2$ s, $t_{run} = 5$ s for 5 krpm; g cm
$t_{acc/dec} = 0.1$ s, $t_{run} = 2.5$ s for 10 krpm; g cm
I^2R = Resistive loss calculated from T_{rms} and K_m; W
P_{total} = total dissipation = $I^2R + P_{core}$; W
(Friction torque was assumed to be 720 g cm)

Note how for the first three motors (A, B, and C), A has the lowest core loss for both profiles and C has the lowest copper loss for both profiles But motor B has the lowest <u>total</u> dissipation; 25% lower than either A or C, and was chosen for this application.

Also, if only the copper loss had been considered, which is usually the case when selecting a motor for typical operation in the 2000 to 3000 rpm range, then the D motor would have been chosen, which would have resulted in the highest dissipation and temperature rise once hardware testing was performed.

Appendix

A.1 Brushless Motor Speed/Torque Curves

The Web Site contains an Excel program that allows the speed/torque curve to be plotted for any brushless motor as a function of:

- Voltage constant: K_e
- Terminal resistance: R_{tt}
- Number of poles
- Terminal inductance: L_{tt}
- Bus voltage
- Thermal resistance
- Core loss
- Ratio of peak torque to continuous torque
- Ambient temperature.

These data will typically be available on the product data sheet, with the exception of the thermal resistance and the core loss factor

A.1.1 Thermal Resistance

This is not always given as a specific value, but implied in the maximum temperature rating for the motor. When not given, use the following approximate values for the closest size motor under consideration, conservatively using the value for the next smaller motor.

Diameter (mm)	Length (mm)	Thermal Resistance ($^{\circ}$C W^{-1})
60	160–200	2
90	190–300	1
120	200–300	0.6
150	300–400	0.5

Electromechanical Motion Systems: Design and Simulation, First Edition. Frederick G. Moritz.
© 2014 John Wiley & Sons, Ltd. Published 2014 by John Wiley & Sons, Ltd.
Companion Website: www.wiley.com/go/moritz

If the thermal resistance is given, correlate the test conditions and mounting plate size used to determine the thermal resistance with the actual mounting to be used in the application

A.1.2 Core Losses

The core loss factor is rarely given. It is determined by both the hysteresis and eddy current effects and is a function of a number of factors, such as pole count, lamination thickness and resistance, lamination annealing, and so on. As such, it is difficult to calculate and measure. However, it does affect the high speed performance of the motor. This program uses the following nominal relation to determine the core loss at any speed N (rpm).

$$P_{CORE} @ N = (P_{CORE} @ 1000 \text{ rpm})(N/1000)^{1.5}$$

When not given, the following shows a range of core loss factors that could be used for a range of motor sizes.

Diameter (mm)	Length (mm)	Core Loss (W @ 1000 rpm)
60	160–200	1–2
90	190–300	2–5
120	200–300	7–15
150	300–400	25–50

Compare Figure A.1 to Figure A.2. The core loss factor in Figure A.2 has been reduced to zero to show the effect as velocity increases from zero along the separation between the continuous and intermittent areas.

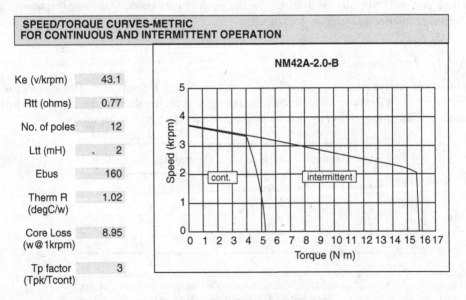

SPEED/TORQUE CURVES-METRIC
FOR CONTINUOUS AND INTERMITTENT OPERATION

NM42A-2.0-B

Ke (v/krpm)	43.1
Rtt (ohms)	0.77
No. of poles	12
Ltt (mH)	2
Ebus	160
Therm R (degC/w)	1.02
Core Loss (w@1krpm)	8.95
Tp factor (Tpk/Tcont)	3

NOTE: MAXIMUM VOLT. LIMITED NO LOAD SPEED (krpm) = 3.71 Tamb = 25

Figure A.1 Speed/torque curve; core losses included (available at www.wiley.com/go/moritz)

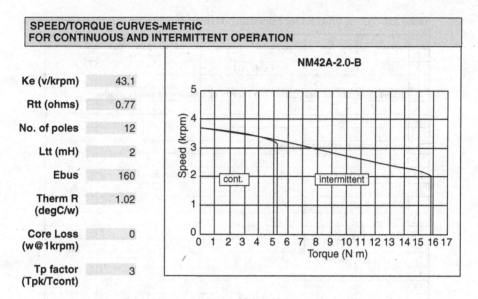

Figure A.2 Speed/torque curve; core losses omitted (available at www.wiley.com/go/moritz)

Note how, in this example, operation at 3000 rpm and 5 N m is inside the continuous region if core losses are not included but becomes an intermittent condition if the core losses are included.

A.2   Inertia Calculation – Excel Program

Cylinders, either solid or hollow, tend to be the predominant shape involved in motion system design and simulation. As such, it is convenient to have an Excel program available to calculate the inertia of cylinders.

Table A.1 shows the Excel program for calculating such inertias along with a list of the densities of materials usually used in motion system components.

Table A.2 shows the result of using this program to calculate the inertia of a hollow steel cylinder.

A.3   Time Constants versus Viscous Damping Constant

In Section 4.2, Equation 4.2, repeated here, shows the exact and approximate characteristic equation for a voltage driven motor.

$$s^2 + \frac{(JR + BL)}{JL}s + \frac{(K_T K_V + BR)}{JL} \approx \left(s + \frac{R}{L}\right)\left(s + \frac{K_V K_T}{JR}\right) \tag{A.1}$$

Note that the approximate expression does not contain the viscous damping constant, B.

Table A.1 Excel Program for calculating the Inertia of a hollow or solid cylinder

	A	B	C	D	E	F	G
1							
2			Table A.1				
3							
4	Excel Program for calculating the Inertia of a hollow or solid cylinder						
5							
6							
7							
8	CYLINDER DIMENSIONS						
9							
10	Inner Radius (cm)		R_i	(for solid cylinder, $R_i = 0$)			
11							
12	Outer Radius (cm)		R_o				
13							
14	Length (cm)		L				
15							
16	Density (gm/cm³)		d				
17							
18							
19		INERTIA =	+((3.1416*C14*C16)/1961.2)*(C12⁴-C10⁴)				
20							
21							
22		MATERIAL DENSITIES					
23							
24		Aluminum	2.66				
25		Brass	8.30				
26		Bronze	8.17				
27		Copper	8.91				
28		Plastic	1.11				
29		Steel	7.75				
30		Wood	0.80				

The term $\frac{R}{L} = \frac{1}{TC_E}$, where TC_E is the electrical time constant and $\frac{K_V K_T}{JR} = \frac{1}{TC_M}$, where TC_M is the mechanical time constant.

If the approximate expression is multiplied out, the following results:

$$s^2 + \frac{(JR + BL)}{JL}s + \frac{(K_T K_V + BR)}{JL} \approx s^2 + \left(\frac{R}{L} + \frac{K_T K_V}{JR}\right)s + \frac{K_T K_V}{JL} \qquad (A.2)$$

If $B = 0$, or if B is "small" such that $BL \ll JR$ and $BR \ll K_T K_E$, then:

$$s^2 + \frac{R}{L}s + \frac{(K_T K_V)}{JL} \approx s^2 + \left(\frac{R}{L} + \frac{K_T K_V}{JR}\right)s + \frac{K_T K_V}{JL} \qquad (A.3)$$

In addition, if $\frac{R}{L} \gg \frac{K_T K_E}{JR}$, then:

$$s^2 + \frac{R}{L}s + \frac{(K_T K_V)}{JL} = s^2 + \frac{R}{L}s + \frac{K_T K_V}{JL} \qquad (A.4)$$

Table A.2 Example for a steel tube, 15 cm long with 5 cm inner radius, 6 cm outer radius

	A	B	C	D	E	F
1						
2			Table A.2			
3						
4		Example for a steel tube, 15 cm long with 5 cm inner				
5		radius, 6 cm outer radius				
6						
7						
8	CYLINDER DIMENSIONS					
9						
10	Inner Radius (cm)		5			
11						
12	Outer Radius (cm)		6			
13						
14	Length (cm)		15			
15						
16	Density (gm/cm^{-3})		7.75			
17						
18						
19		INERTIA =	124.95	g cm s^2		
20						
21						
22	MATERIAL DENSITIES					
23						
24		Aluminum	2.66			
25		Brass	8.30			
26		Bronze	8.17			
27		Copper	8.91			
28		Plastic	1.11			
29		Steel	7.75			
30		Wood	0.80			

Summary: In a second order system in which the electrical time constant is much smaller than the mechanical time constant (which is usually the case) and in which the viscous damping constant is "small", the approximate expression can be used.

If these approximations hold, then:

$$\omega_n^2 = \frac{K_T K_V}{JL} \qquad \omega_n = \sqrt{\frac{K_T K_V}{JL}} \qquad 2\zeta\omega_n = \frac{R}{L} \qquad \zeta = \frac{R}{2}\sqrt{\frac{J}{LK_T K_V}} \qquad (A.5)$$

Note also, in the exact expression, the coefficient of s, $\left(\frac{JR+BL}{JL}\right)$ is defined as $2\zeta\omega_n$ resulting in ζ, the system damping factor being a function of B, the system damping constant for the conditions when $BL \ll JR$ does not hold.

A.4 Current Drive Review

In analyzing a closed loop system in which the motor is powered by an amplifier configured as a current source, it is often convenient and expeditious to model the amplifier simply as a gain block with a transfer function of A V^{-1}. This is warranted since in most systems the amplifier

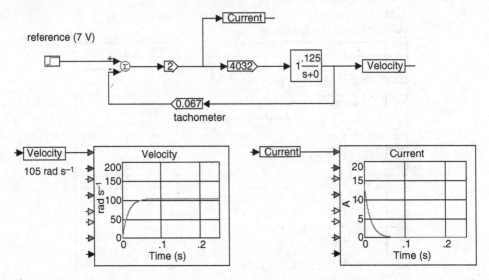

Figure A.3 Basic ideal velocity system; step response, $B = 0$

bandwidth is orders of magnitude higher than the system bandwidth, and the amplifier phase shift will have no or only a minor impact on the system stability.

One problem with this approach is that it does not show any of the responses of the amplifier and does not show how changes in the amplifier parameters affect system operation.

The following Figures A.3–A.10 illustrate a closed loop velocity control system with successively more detailed simulations.

Figure A.3 shows the basic system with an ideal amplifier with a gain of 2 A V^{-1} and a tachometer feedback with a 7 V/1000 rpm calibration, responding to a step command to 1000 rpm (105 rad s^{-1}). This basic system is purely inertial

$$\text{Motor Data} \quad R = 2\,\Omega \qquad K_T = 4032\text{ g cm A}^{-1} \qquad J = 8\text{ g cm s}^2$$
$$L = 20\text{ mH} \quad K_V = 0.40\text{ V rad}^{-1}\text{ s}^{-1} \quad B = 14.4\text{ g cm rad}^{-1}\text{ s}^{-1}$$

Note that in this basic simulation, the system is stable, reaches speed within 0.1 s and experiences an initial peak acceleration current of 14 A. There is no indication as to what should be the current and voltage rating of the amplifier other than 14 A peak provides successful operation. Since the system is purely inertial, steady-state current is zero. Voltage cannot be determined from this simulation.

In Figure A.4, the system has been modified by adding a current limit of $\pm$ 10 A to the amplifier. Performance is still satisfactory and peak current has now been lowered to 10 A.

In Figure A.5, the damping term (B) has been added. Since this term creates a steady-state load, an error is required to support it, resulting in a final velocity of 102 rad s^{-1} (974 rpm) and a steady-state current of 0.36 A.

In Figure A.6, PI compensation has been added to reduce the steady-state error while achieving the final velocity of 105 rad s^{-1} (1000 rpm) and the steady-state current of 0.36 A. The additional P gain results in the final velocity being reached within 0.05 s.

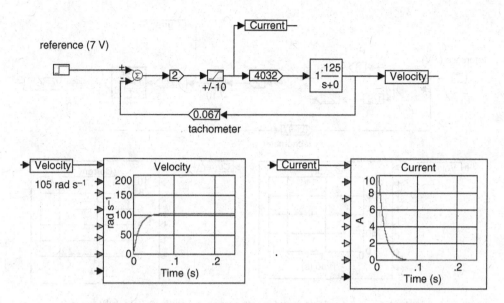

Figure A.4 Basic velocity system, step response, I limit, $B = 0$

Note that the simulation up to this point still provides no information about amplifier voltage.

In Figure A.7, the complete motor is modeled, showing the motor electrical transfer function and the BEMF path. The motor is driven by a current amplifier consisting of a voltage amplifier with a gain of 100 V/V within a current feedback loop with a calibration of 0.48 V A^{-1} resulting in a DC current gain of 2 A V^{-1}, the same gain as in the first four simulations.

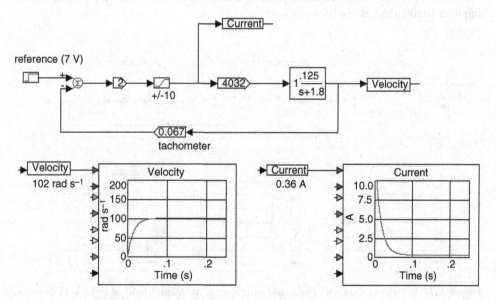

Figure A.5 Basic velocity system, step response, I limit, $B = 14.4$

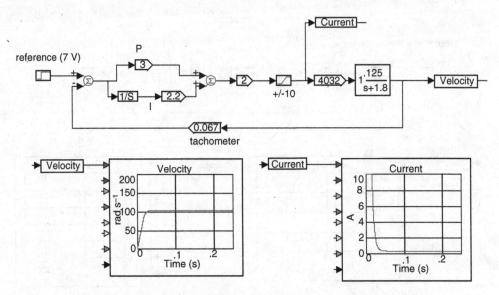

Figure A.6   Basic velocity system, step response, I limit, $B = 14.4$, PI comp

This simulation now displays the output voltage of the amplifier during the transient and steady-state conditions. Note that the amplifier shows a peak output voltage of 500 V at the beginning of the acceleration period. Since it is anticipated that the system will be operated from a 120 V AC power source, creating a 160 V DC supply, the next figure, Figure A.8 has ±160 V limit placed on the amplifier output. System performance remains virtually unchanged.

Up to this point, although current and voltage limits have been placed on the amplifier, the amplifier bandwidth has not been considered.

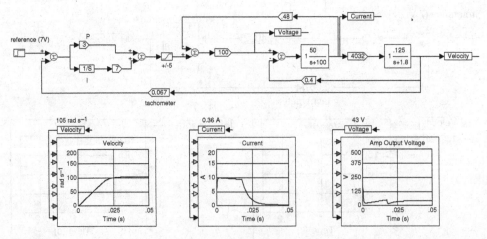

Figure A.7   Velocity system with motor and current amp. modeled, I limit, $B = 14.4$, PI comp, step response

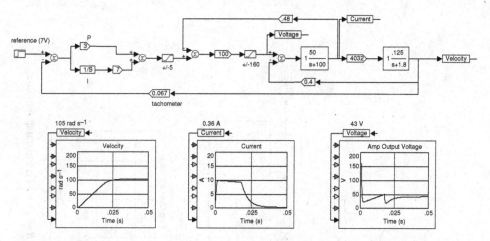

Figure A.8   Velocity system with motor and current amp. modeled, I limit, E limit, $B = 14.4$, PI comp, step response

Figure A.9 shows the amplifier modified to include a double break point at 12 000 rad s^{-1} (2000 Hz) to demonstrate the effect of limiting the amplifier bandwidth to an arbitrarily achievable value. System performance is still satisfactory.

In the light of the currents and voltages displayed and the fact that steady state velocity can be achieved within 0.05 s, it is of interest to demonstrate performance in response to a ramp command rather than the step command.

Figure A.10 shows the simulation of a ramp reference, rising from 0 to 7 V in 0.05 s. The amplifier now shows a peak current of less than 5 A and a peak voltage of less than 50 V.

An amplifier rated at 160 V with a peak current rating of 5 A and a continuous current rating of 0.5 A will be satisfactory for this application.

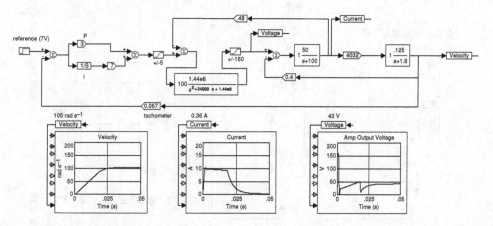

Figure A.9   Velocity system with motor and current amp. modeled, I limit, E limit, amp BW defined, $B = 14.4$, PI comp, step response

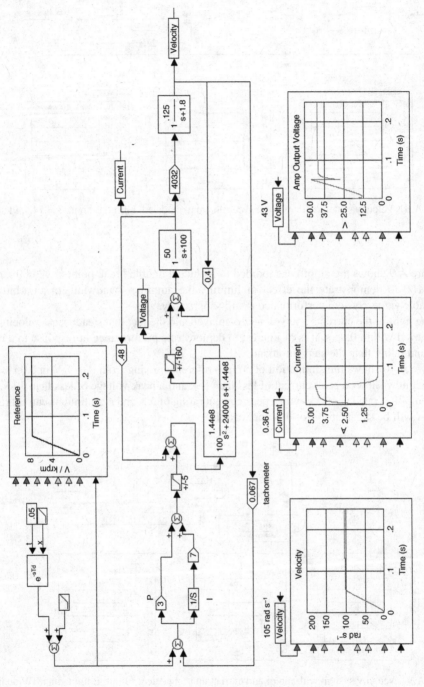

Figure A.10 Velocity system with motor and current amp. modeled, I limit, E limit, amp BW defined, $B = 14.4$, PI comp, ramp response

A.5   Conversion Factors

Length:

$$1 \text{ m} = 100 \text{ cm} = 1000 \text{ mm}$$
$$1 \text{ m} = 39.369 \text{ in}$$
$$1 \text{ cm} = 0.3937 \text{ in}$$
$$1 \text{ in} = 2.54 \text{ cm}$$
$$1 \text{ micron} = 1 \text{ } \mu\text{m}$$

Angular measure:

$$1 \text{ rev} = 360° = 2.16 \times 10^4 \text{ arc min} = 1.296 \times 10^6 \text{ arc s}$$
$$1 \text{ rev} = 2\pi \text{ rad}$$
$$1 \text{ rad} = 0.1592 \text{ rev} = 57.296°$$
$$1 \text{ rad} = 3.438 \times 10^3 \text{ arc min} = 2.063 \times 10^5 \text{ arc s}$$

Linear velocity:

$$1 \text{ m s}^{-1} = 39.369 \text{ in s}^{-1}$$
$$1 \text{ cm s}^{-1} = 0.3937 \text{ in s}^{-1}$$
$$1 \text{ in s}^{-1} = 0.0254 \text{ m s}^{-1} = 2.54 \text{ cm s}^{-1}$$

Angular velocity:

$$1 \text{ rad s}^{-1} = 57.296° \text{ s}^{-1}$$
$$1 \text{ rev min}^{-1} = \frac{1}{60}\text{rev s}^{-1} = \frac{2\pi}{60} \text{ rad s}^{-1}$$
$$1 \text{ rev s}^{-1} = 2\pi \text{ rad s}^{-1}$$
$$1 \text{ rad s}^{-1} = \frac{1}{2\pi}\text{rev s}^{-1}$$

Force/weight:

$$1 \text{ N} = 3.597 \text{ oz} \quad 1 \text{ oz} = 0.278 \text{ N}$$
$$1 \text{ N} = 102 \text{ g} \quad 1 \text{ g} = 9.804 \times 10^{-3} \text{ N}$$
$$1 \text{ g} = 0.0353 \text{ oz} \quad 1 \text{ oz} = 28.35 \text{ g}$$

Torque:

$$1 \text{ N m} = 141.61 \text{ oz in} \quad 1 \text{ oz in} = 0.007\,062 \text{ N m}$$
$$1 \text{ N m} = 0.7376 \text{ lb ft} \quad 1 \text{ lb ft} = 1.356 \text{ N m}$$
$$1 \text{ N m} = 10\,200 \text{ g cm.} \quad 1 \text{ gm cm.} = 9.804 \times 10^{-5} \text{ N m}$$
$$1 \text{ g cm} = 0.013\,89 \text{ oz in} \quad 1 \text{ oz in} = 72.01 \text{ g cm}$$

Inertia:

$$1 \text{ g cm s}^2 = 0.013\,89 \text{ oz in s}^2 \quad 1 \text{ g cm}^2 = 0.001\,02 \text{ g cm s}^2$$
$$1 \text{ oz in s}^2 = 71.98 \text{ g cm s}^2 \quad\quad 1 \text{ oz in}^2 = 0.002\,59 \text{ oz in s}^2$$

Power:

$$1 \text{ HP} = 746 \text{ W} = 550 \frac{\text{ft lb}}{\text{s}} = 746 \frac{\text{N m}}{\text{s}} = 7.607 \times 10^6 \frac{\text{g cm}}{\text{s}}$$
$$1 \text{ kW} = 1.34 \text{ HP} = 1000 \text{ N m s}^{-1}$$

Constants:

$$g = 9.806\,65 \frac{\text{m}}{\text{s}^2} = 980.6 \frac{\text{cm}}{\text{s}^2} = 32.16 \frac{\text{ft}}{\text{s}^2}$$
$$\pi = 3.141\,592\,653\,589\,793\,2$$
$$\lambda = 299\,792\,458 \frac{\text{m}}{\text{s}} = 9.836 \times 10^8 \frac{\text{ft}}{\text{s}}$$

A.6   Work and Power

A.6.1   Work (Energy)

$$\begin{aligned} \text{Work} &= \text{Force} \times \text{Distance} \\ &= \text{N} \times \text{m} = \text{N m} \\ &= \text{g} \times \text{cm} = \text{g cm} \\ &= \text{lb} \times \text{ft} = \text{lb ft} \\ &= \text{oz} \times \text{in} = \text{oz in} \end{aligned}$$

A.6.2   Power

$$\begin{aligned} \text{Power} &= \text{Work per Time} \\ &= \text{N m s}^{-1} \quad &\text{N m min}^{-1} \\ &= \text{g cm s}^{-1} \quad &\text{g cm min} \\ &= \text{lb ft s}^{-1} \quad &\text{lb ft min}^{-1} \\ &= \text{oz in s}^{-1} \quad &\text{oz in min}^{-1} \end{aligned}$$

A.6.3   Horsepower (HP)

Definition:

$$1 \text{ HP} = 550 \text{ lb ft s}^{-1} = 746 \text{ W} = 746 \text{ N m s}^{-1}$$
$$= 33\,000 \text{ lb ft min}^{-1}$$

A.6.4   Rotary Power

$$\text{Distance} = 2\pi R \quad \text{Force} = F$$

$$\text{Work} = 2\pi RF$$

$$\text{Power} = 2\pi RF \times n \left(\frac{\text{rev}}{\text{min}}\right) = 2\pi RFn = (2\pi n)(RF) = 2\pi nT$$

$$(T = RF = \text{Torque})$$

If T is in ft lb and n is in rpm,

$$\text{Then Power} = 2\pi nT \text{ ft lb min}^{-1}$$

$$\text{Since } 1 \text{ HP} = 33\,000 \text{ ft lb min}^{-1}$$

$$\text{HP} = \frac{2\pi nT}{33\,000} = \frac{nT}{5252} \quad T \text{ in ft lb; } n \text{ in rpm}$$

$$= \frac{nT}{1.0084 \times 10^6} \quad T \text{ in oz in; } n \text{ in rpm}$$

$$= \frac{nT}{7120} \quad T \text{ in N m; } n \text{ in rpm}$$

$$= \frac{nT}{7.26 \times 10^7} \quad T \text{ in g cm; } n \text{ in rpm}$$

A.7 I^2R Losses

The following reviews the method for calculating the resistive losses in three types of DC servo motors:

- The conventional, two wire DC motor with wound rotor
- The three phase brushless DC motor with six step drive
- The three phase brushless DC motor with sine wave drive.

In all of the following, it is assumed that the resistance R will be the final value as determined by thermal considerations, as discussed in Section 3.1, Motors and Amplifiers.

A.7.1   Conventional DC Motor

The conventional wound rotor DC motor has two power wires and a fairly constant terminal resistance created by a relatively large number of commutation bars and a number of series/parallel windings. The result is that at any load determined by evaluating load torque and duty cycle, the resistive loss will be:

$$P_R = I_{RMS}^2 R \tag{A.6}$$

R can be determined from the product data sheet or by holding the motor in stall, applying a voltage to the rotor and measuring the current in a number of rotor positions to obtain an average value. Subtract approximately 0.5 V from the applied voltage to account for brush drop.

A.7.2 Three Phase Brushless DC Motor with Trapezoidal BEMF and Six Step Drive

A brushless motor will have either a Wye or Delta connected winding, as shown in Figure A.11.

Each phase is shown as a resistance in series with a BEMF. Inductance is not shown since it does not contribute to dissipation. Input line current is I_{LL} and the input line to line voltage is E_{LL}.

In a six step drive, Wye connection, each phase conducts a "DC" (constant value) current, $I_{LL} = I_P$, for four of the six steps making up a complete commutation cycle.

Therefore, the RMS current for each phase is:

$$I_{RMS} = \sqrt{\frac{2}{3}}\, I_P \tag{A.7}$$

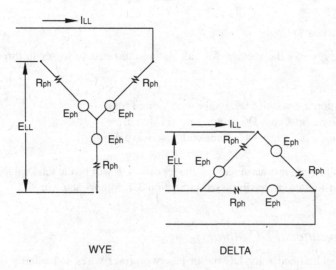

Figure A.11 Brushless Wye/Delta schematics

The total dissipation will be:

$$P_R = \left(\sqrt{\frac{2}{3}}\, I_P\right)^2 2R_{PH} = \frac{4}{3} I_P^2 R_{PH} \tag{A.8}$$

Data sheets usually give resistance as line to line, R_{LL}. Therefore

$$R_{PH} = \frac{R_{LL}}{2} \tag{A.9}$$

and

$$P_R = \frac{2}{3} I_P^2 R_{LL} \tag{A.10}$$

A.7.3 Three Phase Brushless DC Motor with Sine Wave BEMF and Drive

For a sine wave drive, all three phases are continually being supplied with sine wave currents that are 120 electrical degrees apart. As such, conventional three phase calculations hold.

For this case, I_P is the *peak* of the sine wave, therefore, for the Wye connection:

$$I_{RMS} = \frac{I_P}{\sqrt{2}} \tag{A.11}$$

$$P_{PH} = I_{RMS}^2 R_{PH} = \left(\frac{I_P}{\sqrt{2}}\right)^2 \left(\frac{R_{LL}}{2}\right) = \frac{I_P^2 R_{LL}}{4} \tag{A.12}$$

The total dissipation will be three times that of a single phase, or:

$$P_R = \frac{3}{4} I_P^2 R_{LL} \tag{A.13}$$

For the Delta connection:

$$R_{LL} = \frac{(R_{PH})(2R_{PH})}{3R_{PH}} = \frac{2R_{PH}}{3} \tag{A.14}$$

therefore

$$R_{PH} = \frac{3R_{LL}}{2} \tag{A.15}$$

$$P_{PH} = \left(\left(\frac{I_P}{\sqrt{3}}\right)\left(\frac{1}{\sqrt{2}}\right)\right)^2 \left(\frac{3R_{LL}}{2}\right) = \frac{I_P^2 R_{LL}}{4} \tag{A.16}$$

Again, total dissipation will be three times that of a single phase, or:

$$P_R = \frac{3}{4} I_P^2 R_{LL} \tag{A.13}$$

A.8   Copper Resistivity

The resistance of any copper winding at a particular temperature can be determined if the resistance at some base temperature and the absolute temperature is known, which for copper is $-234.5\,°C$.

It is also assumed that temperatures in the region of interest have a linear relationship, as shown in Figure A.12.

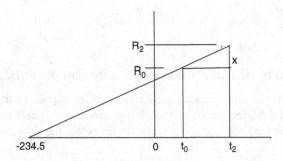

Figure A.12   Copper resistance versus temperature

For typical copper windings, the base temperatures are usually either 20 or 25 °C and the elevated temperatures are either 130 or 155 °C, as determined by the grade of insulation, B or C, respectively.

Example: If $t_0 = 20\,°C$

$$\text{Slope} = \frac{R_0}{234.5 + 20} = 0.003\,93\,R_0 = \frac{x}{t_2 - t_0} \tag{A.17}$$

$$R_2 = R_0 + x = R_0 + 0.003\,93\,R_0\,(t_2 - t_0) \tag{A.18}$$

$$R_2 = R_0\,[1 + 0.003\,93\,(t_2 - t_0)] \tag{A.19}$$

$$\text{If } t_2 = 155\,°C, \text{ then } R_2 = 1.53 R_0 \tag{A.20}$$

$$\text{If } t_2 = 130\,°C, \text{ then } R_2 = 1.43 R_0 \tag{A.21}$$

Similarly, if $t_0 = 25\,°C$, then:

$$R_2 = R_0\,[1 + 0.003\,85\,(t_2 - t_0)] \tag{A.22}$$

$$\text{If } t_2 = 155\,°C, \text{ then } R_2 = 1.50 R_0 \tag{A.23}$$

$$\text{If } t_2 = 130\,°C, \text{ then } R_2 = 1.40 R_0 \tag{A.24}$$

Index

Electromechanical Motion Systems: Design and Simulation, First Edition. Frederick G. Moritz.
© 2014 John Wiley & Sons, Ltd. Published 2014 by John Wiley & Sons, Ltd.
Companion Website: www.wiley.com/go/moritz

Printed in the United States
by Bookmasters

Printed in the United States
By Bookmasters